Springer Proceedings in Energy

The series Springer Proceedings in Energy covers a broad range of multidisciplinary subjects in those research fields closely related to present and future forms of energy as a resource for human societies. Typically based on material presented at conferences, workshops and similar scientific meetings, volumes published in this series will constitute comprehensive state-of-the-art references on energy-related science and technology studies. The subjects of these conferences will fall typically within these broad categories:

- Energy Efficiency
- Fossil Fuels
- Nuclear Energy
- Policy, Economics, Management & Transport
- Renewable and Green Energy
- Systems, Storage and Harvesting
- Materials for Energy

eBook Volumes in the Springer Proceedings in Energy will be available online in the world's most extensive eBook collection, as part of the Springer Energy eBook Collection. To submit a proposal or for further inquiries, please contact the Springer Editor in your region:

Kamiya Khatter (India)
Email: kamiya.khatter@springer.com
Loyola D'Silva (All other countries)
Email: loyola.dsilva@springer.com

Chinnathan Areeprasert · Thongthai Witoon ·
Gasidit Panomsuwan · Chanoknunt Khaobang ·
Wen Da Oh
Editors

Proceedings of the International Conference on Sustainable Energy and Environmental Technology for Circular Economy

SEECE 2024; 04–06 Aug, Bangkok, Thailand

Springer

Editors
Chinnathan Areeprasert
Faculty of Engineering
Kasetsart University
Bangkok, Thailand

Thongthai Witoon
Faculty of Engineering
Kasetsart University
Bangkok, Thailand

Gasidit Panomsuwan
Faculty of Engineering
Kasetsart University
Bangkok, Thailand

Chanoknunt Khaobang
Faculty of Engineering
Kasetsart University
Bangkok, Thailand

Wen Da Oh
School of Chemical Sciences
Universiti Sains Malaysia
Pulau Pinang, Malaysia

ISSN 2352-2534 ISSN 2352-2542 (electronic)
Springer Proceedings in Energy
ISBN 978-981-96-2305-1 ISBN 978-981-96-2306-8 (eBook)
https://doi.org/10.1007/978-981-96-2306-8

This Springer imprint is published by the registered company Springer Nature Singapore Pte Ltd.
The registered company address is: 152 Beach Road, #21-01/04 Gateway East, Singapore 189721, Singapore

Organization

Editor

Chinnathan Areeprasert, Kasetsart University, Thailand
Chanoknunt Khaobang, Kasetsart University, Thailand
Gasidit Panomsuwan, Kasetsart University, Thailand
Thongthai Witoon, Kasetsart University, Thailand
Wen Da Oh, Universiti Sains Malaysia, Malaysia

Technical Chair

Chinnathan Areeprasert, Kasetsart University, Thailand
Chanoknunt Khaobang, Kasetsart University, Thailand
Gasidit Panomsuwan, Kasetsart University, Thailand
Thongthai Witoon, Kasetsart University, Thailand
Wen Da Oh, Universiti Sains Malaysia, Malaysia

Technical Committee

Anil Ghule, Shivaji University, India
Chaiyaporn Silawatchananai, King Mongkut's University of Technology North Bangkok, Thailand
Chan Yen San Stephanie, Curtin University, Malaysia
Chart Chiemchasri, Kasetsart University, Thailand
Chootrakul Siripaiboon, Sukhothai Thammathirat Open University, Thailand
Dachao Ma, Guangxi University, China
Her-Terng Yau, National Chung Cheng University, Taiwan

Hui Lin Ong, Universiti Malaysia Perlis, Malaysia
Kentaro Umeki, Lulea University of Technology, Sweden
Kreetha Somkeattikul, Panyapiwat Institute of Management, Thailand
Lu Ding, East China University of Science and Technology, China
Metta Chareonpanich, Kasetsart University, Thailand
Mi Yan, Zhejiang University of Technology, China
Nakorn Tippayawong, Chiang Mai University, Thailand
Naragain Phumchusri, Chulalongkorn University, Thailand
Narayan Babu Dhita, Patan Multiple Campus Tribhuvan University, Nepal
Navadol Laosiripojana, King Mongkut's University of Technology Thonburi, Thailand
Pandji Prawisudha, Bandung Institute of Technology, Indonesia
Papis Wongchaisuwat, Kasetsart University, Thailand
Phantisa Limleamthong, Kasetsart University, Thailand
Remya Neelancherry, Indian Institute of Technology Bhubaneswar, India
Tanakorn Wongwuttanasatian, Khon Kaen University, Thailand
Yong-Song Chen, National Chung Cheng University, Taiwan
Zhao Peitao, China University of Mining and Technology, China

Preface

The 2024 International Conference on Sustainable Energy and Environmental Technology for Circular Economy (SEECE 2024) was organized from 4 to 6 August 2024 by Kasetsart University, Bangkok, Thailand.

This year, *Innovative Pathways for Sustainable Energy and Environment Towards a Circular Economy*, was selected as the conference theme, aiming to provide an academic and professional platform for discussing groundbreaking research across diverse topics, including but not limited to:

- Biomass and Waste Conversion
- Material from Biomass and Waste
- Renewable Energy Technology
- Analysis of Energy and Environment Systems

SEECE 2024 marks the first installment of this conference, organized by the Faculty of Engineering and Kasetsart University Research and Development Institute in collaboration with other international partners. The 2-day hybrid conference (online and in-person) welcomed *48 presenters*, with a total of *18 papers accepted* after a peer-review process to ensure that all research aligns with the theme and objectives of SEECE 2024 for publication in conference proceeding.

The editors extend their heartfelt appreciation to all contributors for their invaluable insights shared during SEECE 2024. We are also deeply grateful to the keynote speakers, peer reviewers, and organizing committee members for their dedication to making this conference a resounding success. Special thanks are given to Springer for their support in producing these proceedings.

Bangkok, Thailand — Chinnathan Areeprasert
Bangkok, Thailand — Thongthai Witoon
Bangkok, Thailand — Gasidit Panomsuwan
Bangkok, Thailand — Chanoknunt Khaobang
Pulau Pinang, Malaysia — Wen Da Oh

Contents

Part I
Biomass and Waste Conversion

Chapter 1
An Investigation on the Efficiency of the Carbonization Process in Beachside Waste Management

Tanaporn Thapsamut and Chinnathan Areeprasert

Abstract This study investigates the carbonization process of beachside waste utilizing a community-scale carbonizer. The waste samples were collected from Bang Saen Beach, located in Chonburi Province on the east coast of Thailand, which is a popular tourist destination from the results of the research, it was found that a total of 16 categories of waste could be separated. The three primary types of waste identified are plastic, small other wastes, and wood or organic waste. Their respective proportions are 58.47%, 14.94%, and 14.35%, respectively. The carbonization reactor involves two different chambers: combustion and carbonization. The former generates heat through the combustion of rubber firewood, which is then directed to the carbonization chamber. Exhaust gases are treated using a water scrubber. Carbonization occurs at a temperature of 450 °C with a heating rate of 2.325 °C/min. The primary wet yield of beachside waste char stands at 43.47%, while the calorific value of beachside waste char is 2,176 cal/g.

Keywords Carbonization · Waste management · Coastal debris · Beachside waste biochar

1.1 Introduction

In recent years, beachside waste management poses a significant environmental challenge in tourist destinations worldwide, including Bang Saen Beach, Chonburi Province, Thailand, where plastic, unidentified small waste, and organic debris accumulate [1]. Traditional waste disposal methods like incineration and landfilling, though widely used, present environmental and spatial concerns unsuitable for coastal areas [2, 3]. Biochar production via carbonization offers a promising alternative,

T. Thapsamut · C. Areeprasert (✉)
Department of Mechanical Engineering, Faculty of Engineering, Kasetsart University, 50 Ngam Wong Wan Road, Lat Yao, Chatuchak, Bangkok 10900, Thailand
e-mail: fengcta@ku.ac.th

C. Areeprasert et al. (eds.), *Proceedings of the International Conference on Sustainable Energy and Environmental Technology for Circular Economy*, Springer Proceedings in Energy, https://doi.org/10.1007/978-981-96-2306-8_1

converting organic waste into biochar a stable, carbon-rich material beneficial for soil amendment and carbon sequestration [4–6].

The carbonization process involves heating organic waste in the absence of oxygen, a method used effectively to convert into valuable biochar products. Key parameters of carbonization process are temperature, heating rate, residence time, and type of feedstock. In this work, our in-house developed carbonizer has been utilized. It is consists of two reactors, where biomass fuel such as rubber firewood undergoes combustion in a combustion chamber to generate heat which is sent to a carbonization chamber for carbonization process of preferred feedstock at approximately 450 °C. To meet environmental standards, exhaust gases from combustion are treated using a water scrubber [6]. To convert waste into more valuable material, carbonization could be one of the promising methods that is simple and easy for implementation.

This study contributes significantly to sustainable waste management practices by demonstrating the feasibility of biochar production from beachside waste. By converting waste into biochar, communities not only mitigate environmental impacts associated with waste accumulation but also harness valuable resources for soil improvement and energy generation. Such initiatives highlight the potential for carbonization technologies to play a pivotal role in transforming waste management practices, particularly in coastal regions facing increasing waste challenges.

1.2 Material and Methods

1.2.1 Raw Material

This study was conducted at bangsean beach in Chonburi Province in the East region of Thailand. Bangsaen beach (13° 17′ 55.4″ N 100° 54′ 08.0″ E) is a popular attraction site. It is often bustling with tourists. An experiment for waste collection was conducted to gather samples. A rectangular area measuring (50 × 5 m) was designated as the collection zone between the routes, beginning from the Chao Pho Saen Shrine to the Bangsaen Beach roundabout. A total of fifteen cells situated along the shoreline were sampled for the purpose of collecting and analyzing the composition of the BSW. Collected debris was placed in sealed bags, properly labeled, and stored for further examination. The gathered debris was then sorted into 16 categories, which included plastic, wood/organic, nylon, metal/steel, paper, rubber, sack bags, plastic rope, textiles, glass, smoking debris, sponge, mask, foam, net, and other items. The debris was categorized based on their wet weight proportions as shown in Table 1.1.

Table 1.1 Beachside waste composition without incombustible parts

Composition	Plastic	Organic	Sack bag	Paper	Textile	Foam	Plastic rope
As received basis (wt.%)	58.47	14.35	0.40	1.01	1.75	0.21	0.17
Composition	Smoking debris	Nylon	Rubber	Sponge	Net	Mask	Other
As received basis (wt.%)	0.11	3.90	0.98	0.32	2.90	0.50	14.94

1.2.2 Pilot-Scale Carbonization

In this work, A direct-heating, pilot-scale carbonizer consists of two horizontal chamber reactors that serve as the combustor (the heat generator) and the carbonized (the char generator), respectively. Each reactor possessed dimensions of 800 (Width) × 800 (Length) × 1000 (Depth) mm. It was constructed using 10 mm-thick refractory brick combined and 3 mm-thick SS400 steel. Air was fed into the combustion chamber through a blower with a power of 1 hp. During the carbonization process, the targeted temperature stood at 450 °C, and the duration of the reaction residence was set at 3 h. The carbonization procedure was conducted with a low heating rate of 2.325 °C /min to achieve a high solid yield.

1.2.3 Material Characterization

Full experiment and analyzed process diagram was shown in Fig. 1.1. To assess the physical and chemical characteristics of the beachside char, a sample was first dried at 105 °C in an electric oven for 24 h, then pulverized into powder and stored for subsequent analysis. Proximate analysis was conducted using a simultaneous thermal analyzer (TGA801) following ASTM D7852 standards and X-ray fluorescence (XRF) was carried out for ash composition analysis. The heating value of the char was determined using an automatic bomb calorimeter (Leco model AC-500).

1.3 Results and Discussion

1.3.1 Performance of Pilot-Scale Carbonization

Figure 1.2 shows temperature profiles from a pilot-scale carbonizer across various phases. Initially, the temperature remained steady (100–200 °C) during the first 30 min, reflecting the drying process. Concurrently, the combustion chamber temperature peaked within the initial 2 h, stabilizing around 600 °C due to the use of eucalyptus firewood. This led to temperature variations influenced by the moisture

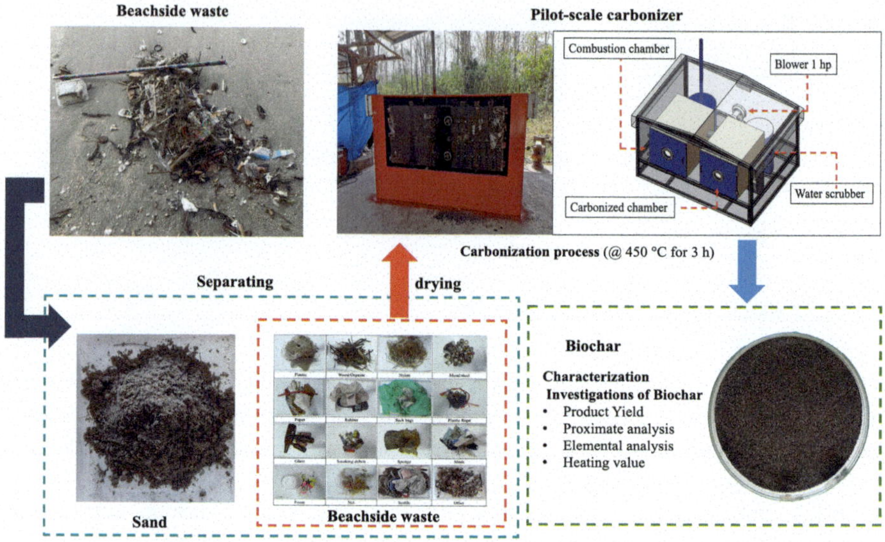

Fig. 1.1 Experiment flowchart and details of work conducted

and density of the wood. Heat transfer during this phase predominantly occurred via exhaust gases from the combustion chamber, converting liquid moisture into vapor. Torrefaction started between 30 and 120 min, marked by a notable temperature increase, resulting in a product with higher energy density but lower fixed carbon content. Carbonization began after 2 h at approximately 450 °C, achieving low-temperature carbonization at a rate of 2.325 °C/min. The extended volatile residence time and slower heating rate promoted equilibrium reactions, enhancing the yield of char.

Eucalyptus wood was used as fuel, which has a heating value of about 4,550 kcal/kg. The heating process took 3 h to reach the target temperature. The actual carbonization process began when the temperature reached around 450 °C, and the carbonization process was controlled at the temperature between 400 and 500 °C. Since the heating rate was 2.325 °C/min, this slow heating rate and longer carbonization temperature allowed to obtain higher char yield [7]. The carbonization of BSW resulted in a solid product yield of 43.47%. However, the BSW contains a lot of volatile components, mainly plastics, it could reduce char yield.

1.3.2 Product Characterization

The proximate analysis results for the BSW char products are provided in Table 1.2. Volatile matter refers to the components of biochar that burn and release gases when heated. A higher volatile matter content indicates a significant amount of organic

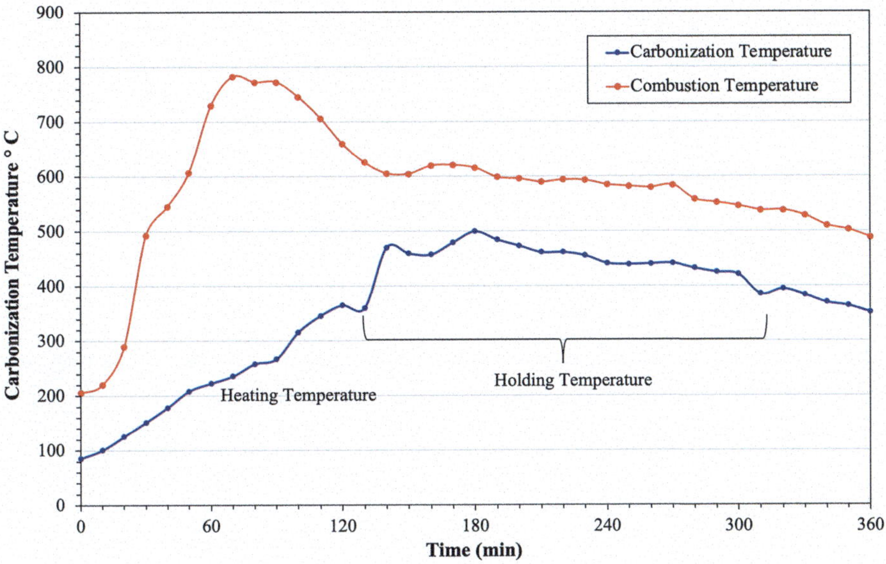

Fig. 1.2 Temperature profile of pilot-scale carbonization

material, possibly from a large quantity of plastic waste. While this high volatile content may be beneficial for certain uses, the biochar's lower fixed carbon and higher ash content could limit its effectiveness as a fuel compared to other biochar types. Fixed carbon is crucial for biochar quality as a fuel, and good biochar typically has a higher fixed carbon content [8].

The compositions of the BSW char products are presented in Table 1.3. From the composition test results, it was found that BSW char has a combined content of approximately 78% silicon dioxide and alumina dioxide, which is enough to be used as a precursor for producing geopolymer. Mengya et al. studied on mechanical and microstructure properties of geopolymer concrete via municipal solid waste (MSW) ash. It was found that MSW ash contains the amount of Si and Al was equal to 75.54%. In addition, the result shows the compressive strength of the geopolymer reached 26.24 MPa when the replacement rate of 20%. MSW ash [9]. This could be

Table 1.2 Proximate analysis of beachside waste biochar

Item	Unit	Value
Proximate analysis		
Moisture	wt.%	4.32
Volatile matter	wt.%	41.9
Ash	wt.%	35.13
Fixed carbon	wt.%	18.65

Table 1.3 Ash composition analysis of beachside waste biochar

Chemical composition (wt.%)									
SiO_2	Al_2O_3	Fe_2O_3	MnO	MgO	CaO	Na_2O	K_2O	TiO_2	P_2O_5
75.4	3.13	0.56	0.0141	0.614	5.83	1.12	2.5	0.0682	0.157
Trace element content (ppm)									
As	Ba	Ti	Cu	Pb	Zn	Rb	Sr	Br	Mn
13.21	223	454	44.6	ND	113	153	182	22.2	92.8

*ND meaning no detects

an interesting way of reutilization of BSW char in alternative construction material application.

1.4 Conclusions

This study underscores the feasibility and benefits of employing carbonization technology for sustainable waste management in coastal areas, offering a pathway towards reducing environmental impact and promoting resource recovery. The result led to the following conclusions:

- Primary waste types identified included plastics, unidentified small other waste, and organic materials, with significant proportions suitable for carbonization.
- The pilot-scale carbonization process yielded beachside waste char with a substantial fixed carbon content of 34.65% and a wet yield of 43.47%, indicating efficient conversion rates under controlled conditions.
- BSW char has a combined content of approximately 78% silicon dioxide and alumina dioxide, which is enough to be used as a precursor for pozzolanic material.
- The char produced exhibited a high calorific value of 2,176 cal/g, its potential as an efficient energy source.

Acknowledgements This research is supported by the Kasetsart University Research Development Institute – FF(KU)51.67. T. Thapsamut received a scholarship from the Faculty of Engineering at Kasetsart University.

References

1. J. Dąbrowska, M. Sobota, M. Świąder, P. Borowski, A. Moryl, R. Stodolak, J.K. Kazak, Marine waste—sources, fate, risks, challenges and research needs. Int. J. Environ. Res. Public Health **18**(2), 433 (2021)

2. C.J. Brown, S.D. Jupiter, S. Albert, K.R. Anthony, R.J. Hamilton, A. Fredston-Hermann, C.J. Klein, A guide to modelling priorities for managing land-based impacts on coastal ecosystems. J. Appl. Ecol. **56**(5), 1106–1116 (2019)
3. C.J. Brown, M.F. Adame, C.A. Buelow, M.A. Frassl, S.Y. Lee, B. Mackey, R.M. Connolly, Opportunities for improving recognition of coastal wetlands in global ecosystem assessment frameworks. Ecol. Ind. **126**, 107694 (2021)
4. A. Kumar, E. Singh, L. Singh, S. Kumar, R. Kumar, Carbon material as a sustainable alternative towards boosting properties of urban soil and fostering plant growth. Sci. Total. Environ. **751**, 141659 (2021)
5. M. Verma, N. M'hamdi, Z. Dkhili, S.K. Brar, K. Misra, Thermochemical transformation of agro-biomass into biochar: simultaneous carbon sequestration and soil amendment, in F. Editor, S. Editor (eds.) *Biotransformation of waste biomass into high value biochemical* 51–70 (2014)
6. Y.X. Seow, Y.H. Tan, N.M. Mubarak, J. Kansedo, M. Khalid, M.L. Ibrahim, M. Ghasemi, A review on biochar production from different biomass wastes by recent carbonization technologies and its sustainable applications. J. Environ. Chem. Eng. **10**(1), 107017 (2022)
7. F. Amalina, A.S. Abd Razak, S. Krishnan, H. Sulaiman, A.W. Zularisam, M. Nasrullah, Biochar production techniques utilizing biomass waste-derived materials and environmental applications—a review. J. Hazard. Mater. Adv. **7**, 100134 (2022)
8. A. Shariff, N.S.M. Aziz, N. Abdullah, Slow pyrolysis of oil palm empty fruit bunches for biochar production and characterization. J. Phys. Sci. **25**(2), 97 (2014)
9. M. Niu, P. Zhang, J. Guo, J. Wang, Effect of municipal solid waste incineration fly ash on the mechanical properties and microstructure of geopolymer concrete. Gels **8**(6), 341 (2022)

Chapter 2
Effect of Biochar and Hydrochar Germination Rate of Khon Kaen 3 Sugarcane Bud Seed Under Terrible Soil Conditions

Kor Taweengern, Surachet Aramrak, and Chinnathan Areeprasert

Abstract This study explores the transformative effects of biochar and hydrochar amendments on the germination rate of Khon Kaen 3 sugarcane bud seeds under harsh conditions of unsustainable soils in sugarcane cultivation. Biochar and hydrochar, derived from bagasse carbonization and hydrothermal carbonization, were applied to unsuitable soils at the mixing rate of 4%, resulting in notable improvements in physical growth and percent brix. Preliminary findings reveal a significant enhancement in the germination rate of Khon Kaen 3 sugarcane bud seeds. Results also showed that the application of biochar and hydrochar can improve soil moisture retention, which directly contributes to higher germination rates and overall plant growth. This research underscores the promising role of biochar and hydrochar in fostering soil resilience and promoting crop establishment in challenging environments.

Keywords Biochar · Hydrochar · Sugarcane bagasse · Soil improvement · Soil amendment · Germination

2.1 Introduction

Brazil is the world leader in term of sugarcane production output, followed by China, India, and Thailand. Brazil produced 642.7 million tons of sugarcane on 8.44 million hectares in 2019–2020 [1]. There were 106 million tons of sugarcane available in Thailand in 2015 [2]. Production is impacted by elements including low soil quality

K. Taweengern · C. Areeprasert (✉)
Department of Mechanical Engineering, Faculty of Engineering, Kasetsart University, Bangkok 10900, Thailand
e-mail: fengcta@ku.ac.th

S. Aramrak
Department of Soil Science, Faculty of Agriculture, Kasetsart University, Bangkok 10900, Thailand

C. Areeprasert et al. (eds.), *Proceedings of the International Conference on Sustainable Energy and Environmental Technology for Circular Economy*, Springer Proceedings in Energy, https://doi.org/10.1007/978-981-96-2306-8_2

and automated harvesting. Global yields are significantly impacted by drought. From 2015 to 2017 [3], Thailand was hit by severe droughts, which caused yield decreases of 11% in 2016 and 12.4% in 2017 [4]. Increasing yields requires enhancing the availability of water, nutrients, and high-quality soil.

For every ton of raw sugarcane, sugar refineries generate about 0.3 tons of bagasse waste [5]. Compost feedstock, biodegradable materials, reinforcing materials, combustion fuel, and soil amendment char are among the uses for bagasse [6]. Utilizing hydrochar and biochar to improve soil has been the subject of recent research. According to Bento et al., adding 4% (w/w) of hydrochar to soil produced a high level of organic carbon and the release of nutrients, suggesting that this material might be used in place of fertilizer [7]. According to Farid et al., co-composted biochar increased seed germination and soil organic carbon [8]. According to Alves et al., biochar made from bagasse and sewage sludge raised soil carbon and increased soil stability [9].

This study evaluated char production via carbonization and hydrothermal carbonization (HTC) from sugarcane bagasse as soil amendments on sugarcane bud seed germination in different types of soil and cultivating conditions. Investigation on the physical properties of sugarcane bud seeds, germination index, physical growth, and Brix energy were conducted to assess the potential of biochar and hydrochar as soil amendments for sugarcane production.

2.2 Material and Methods

2.2.1 Sugarcane Bagasse

The raw material for this study was sugarcane bagasse, which was provided by Mitr Phol Phu Khieo Bio-Power (Mitr Phol Bio-Power Co., Ltd.). Sugarcane bagasse was sun-dried outside for several days at a temperature between 27.5 to 34.5 °C.

2.2.2 Biochar and Hydrochar Production

The lab-scale carbonizer has a diameter of 45 mm and a height of 400 mm. It is constructed of stainless steel (SUS316). The carbonizer's heater has an 8 kW power source. The process temperature is controlled by the PID controller. Seventy grams of the raw materials for each experiment were used. The operational temperatures was 400 °C, with a heating rate of 5 °C per min. The carbonization residence time was 60 min. To prevent biomass combustion process during carbonization, 50 ml/min of nitrogen flow was fed into the reactor. For comparison, a target temperature was maintained for 1 h. The produced biochar has carbon, hydrogen, and nitrogen of 80.4%, 3.7%, and 0.6%, respectively.

The raw material used in the HTC experiment was chopped into tiny pieces, and no additional drying step was carried out. The apparatus used was a 500 ml high-pressure autoclave reactor assembled with a stirrer. 150 ml of distilled water and 25 g of raw sugarcane bagasse were introduced to the reactor chamber for each experiment. The ratio of the sample to distilled water was 1:6, allowing a complete submersion of the sample in the water for an effective reaction. To maintain a uniform temperature distribution within the reactor, the stirrer was maintained at 25 revolutions per min. HTC temperatures were measured and held for 30 min at 160 °C. Throughout the whole experiment, the HTC experiment's heating rate was roughly 4.18 °C/min. The produced hydrochar has 47.4%, 6.0%, and 0.1% of carbon, hydrogen, and nitrogen, respectively.

2.2.3 *Sugarcane Cultivation Soil*

This research tested two soil types for sugarcane cultivation: clay loam and sand, collected from Khon Sawan district, Chaiyaphum province, Thailand. The samples underwentcleaning process using a 2-mm round sieve to remove contaminants. Hydrochar and biochar were added to each soil type at 4% by mass for the subsequent tests.

2.2.4 *Germination Test of Char-Amended Soil in Several Conditions*

Germination tests were conducted on sugarcane buds from Khon Kaen 3, which were cut to lengths of 47 mm and diameters of 28 mm. The buds were soaked in water for 24 h before the experiment. Sand or clay loam was mixed with biochar and hydrochar, with the incubation periods of 14 days for sand and 21 days for clay loam, at a temperature of 27.5 °C. These different incubation periods were due to the physical and chemical effects of the soil textures on incubation speed. The char mixing ratio was 4% by weight, and 15 samples were tested for each condition.

The experiment simulated three growing conditions: normal or control (daily watering), flooding (15 days of normal watering followed by submerging the buds in water), and drought (5 days of normal watering followed by no irrigation). Indicators for the germination tests included germination indices of sett roots, primary leaves, and Brix percentage. Data were collected every 5 days over a 25-day period, with the findings presented as relative percentages. True density is measured by the weight per unit volume of the sugarcane bud seed. Vernier calipers were used for measuring physical growth and moisture content was calculated by the mass difference before and after drying at 105 °C for 24 h.

2.3 Results and Discussion

2.3.1 Physical Properties of Sugarcane Bud Seed

After soaking, the upper, middle, and lower segments of sugarcane bud seeds increased by 0.18–1.03%. The diameter of the bud seeds increased by 0.97%, and their length increased by 0.51%. The weight and true density per chip rose by 4.33% and 3.58%, respectively, and the moisture content increased by 1.31%. These findings provide insight into the expansion of sugarcane cuttings when they are prepared before planting, which is crucial for designing appropriate machinery to support this process.

2.3.2 Germination Index

The sett roots' germination index (GI) is presented in Fig. 2.1a and 2.1b for sand and clay loam, respectively. Three cultivation scenarios drought, flooded, and control were examined. This is a summary of the soil and sugarcane buds' moisture contents during the experiment. Regardless of the growth scenarios and soil amendment conditions, all samples exhibited 100% GI of sett roots after 15 days of germination. It was sugarcane of the Khon Kaen 3 variety. The growth rate of sugarcane is naturally high [10].

The germination index of the primary leaf (GI-PL) is presented in Fig. 2.1c and 2.1d for sand and clay loam, respectively. The original control soils had higher GI-PL values compared to those under severe conditions, indicating the importance of proper watering. Biochar and hydrochar significantly improved GI-PL. In sand under drought conditions, char-amended samples had a GI-PL of 40–45%, higher than that of 35–40% under flooded conditions. In clay loam, biochar and hydrochar resulted in similar GI-PL values, reflecting differences in water and nutrient storage.

2.3.3 Physical Growth

Biochar and hydrochar enhanced the number of sett roots across both soil types under all experimental conditions. In sand soils under drought conditions, root diameter was 50.0% and 62.5% increase from control when adding biochar and hydrochar, respectively. Under the same condition, root length also showed significant improvement when using biochar (112.5%) and hydrochar (220.0%), similarly, in clay loam soils, they promoted root length under all conditions. Additionally, biochar and hydrochar increased stem size in both soil types under drought conditions and promoted stem length across all conditions. The number of leaves also increased in both soil types under drought and normal conditions with biochar and

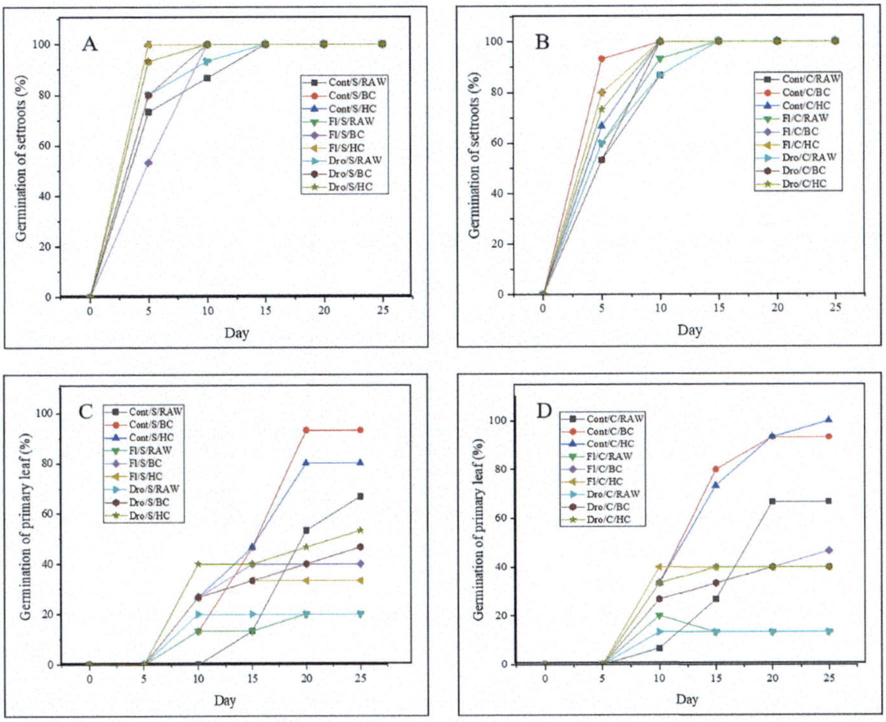

Fig. 2.1 Germination index of sett root **a** sand; **b** clay loam and germination index of the primary leaf **c** sand; **d** clay loam

hydrochar applications. Biochar further improved leaf length, leaf size, and apex length for both soil types in all conditions. Moreover, biochar and hydrochar significantly reduced soil moisture loss in both soil types, as well as seed bud moisture loss under drought conditions. These findings demonstrate that biochar and hydrochar serve as valuable soil amendments, playing a crucial role in enhancing sugarcane growth by improving soil structure and increasing water retention. Additionally, they enhance root efficiency, allowing plants to absorb nutrients more effectively, which results in stronger stems, larger leaves, and significantly increased overall growth. Their ability to maintain soil moisture also supports plant growth under both drought and flooding conditions.

2.3.4 Brix Values

Figure 2.2 present the Brix values and the impact of biochar and hydrochar on sugarcane bud seeds under control, flooded, and drought conditions. Under normal conditions, biochar mixed with sand and clay loam soils significantly reduced Brix

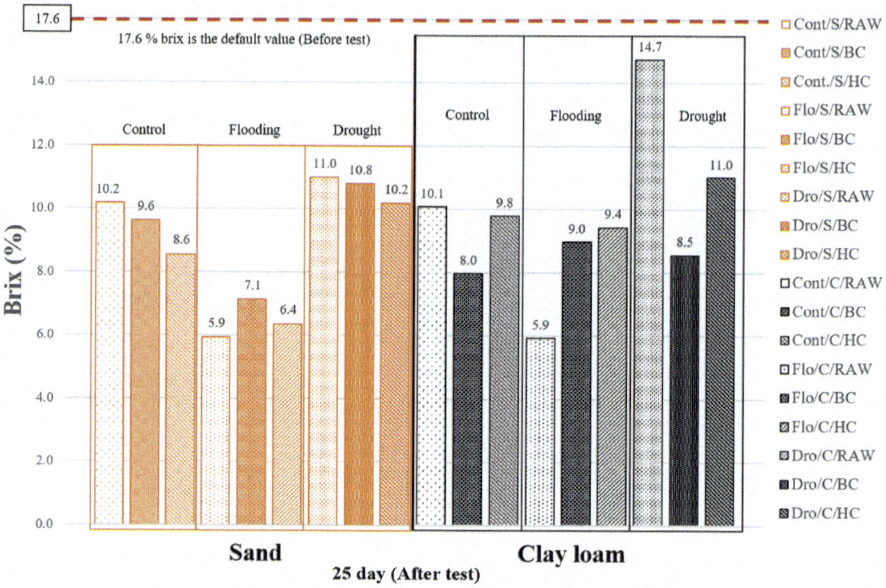

Fig. 2.2 Brix values of sugarcane bud before and after germination tests

values compared to hydrochar and control soils, indicating that biochar aids in energy release, promoting germination. Under flooding conditions, soils without biochar or hydrochar showed a significant drop in Brix values, with flooding stunting germination and soils mixed with biochar and hydrochar notably slowing energy supply. In drought conditions, sand soil mixed with biochar showed a non-significant decrease in Brix values, while hydrochar mixed with sand soil showed a significant decrease. Both biochar and hydrochar mixed with clay loam soil significantly reduced Brix values, correlating with improved growth and moisture retention, supporting sugarcane growth under drought conditions.

2.4 Conclusion

This study investigates the impact of biochar and hydrochar on enhancing the quality of unsuitable soils and improving the germination rate of Khon Kaen 3 sugarcane bud seeds, particularly under harsh environmental conditions such as drought and flooding. The findings reveal that the application of biochar and hydrochar plays a significant role in improving soil moisture retention, which directly contributes to higher germination rates and overall plant growth. Specifically, biochar and hydrochar facilitate the release of energy during the early stages of germination, providing essential support for seedling development and promoting faster and more uniform sprouting. Therefore, the incorporation of biochar or hydrochar for improving sugarcane cultivation in areas where soil conditions are suboptimal could be a promising solution.

Acknowledgements This research was supported by the Kasetsart University Research and Development Institute (KURDI, FF(KU)25.64). K. Taweengern received a scholarship from the Faculty of Engineering, Kasetsart University.

References

1. D.E. Cursi, H.P. Hoffmann, G. Barbosa, J.A. Bressiani, R. Gazaffi, R.G. Chapola et al., History and current status of sugarcane breeding, germplasm development and molecular genetics in Brazil. Sugar Tech. **24**(1), 112–133 (2022)
2. P. Sukyai, N. Yingkamhaeng, N.T. Lam, V. Tangsatianpan, C. Watcharinrat, G. Vanitjinda et al., Research and development prospects for sugarcane and sugar industry in Thailand. Sugar Tech. **18**, 583–587 (2016)
3. G. Leng, J. Hall, Crop yield sensitivity of global major agricultural countries to droughts and the projected changes in the future. Sci. Total. Environ. **654**, 811–821 (2019)
4. C. Apirumanekul, M. Shrestha, Y. Inmoung, R. Ratanavaraha, J. Kaewsuk, Application of robust decision support (RDS) for water scarcity management in Northeast Thailand, *Development and Climate Change in the Mekong Region* 261
5. L.A.R. Batalha, Q. Han, H. Jameel, H-M. Chang, J.L. Colodette, F.J.B. Gomes, Production of fermentable sugars from sugarcane bagasse by enzymatic hydrolysis after autohydrolysis and mechanical refining. Bioresour. Technol. **180**, 97–105 (2015)
6. E. Ajala, J. Ighalo, M. Ajala, A. Adeniyi, A. Ayanshola, Sugarcane bagasse: a biomass sufficiently applied for improving global energy, environment and economic sustainability. Bioresour. Bioprocessing. **8**, 1–25 (2021)
7. L.R. Bento, A.J.R. Castro, A.B. Moreira, O.P. Ferreira, M.C. Bisinoti, C.A. Melo, Release of nutrients and organic carbon in different soil types from hydrochar obtained using sugarcane bagasse and vinasse. Geoderma **334**, 24–32 (2019)
8. A. Gross, T. Bromm, B. Glaser, Soil organic carbon sequestration after biochar application: a global meta-analysis. Agronomy **11**(12), 2474 (2021)
9. B.S.Q. Alves, K.P.S. Zelaya, F. Colen, L.A. Frazão, A. Napoli, S.J. Parikh et al., Effect of sewage sludge and sugarcane bagasse biochar on soil properties and sugar beet production. Pedosphere **31**(4), 572–582 (2021)
10. J. Khonghintaisong, P. Songsri, N. Jongrungklang, Understanding growth rate patterns among different drought resistant sugarcane cultivars during plant and ratoon crops encountered water deficit at early growth stage under natural field conditions. Agronomy **11**(10), 2083 (2021)

Chapter 3
H_2/CO-Rich Syngas Production from Biochar CO_2/O_2 Gasification

Nutthaphon Yutthakarn, Chanoknunt Khaobang, Sasipa Boonyubol, Chinnathan Areeprasert, and Jeffrey S. Cross

Abstract This paper investigated a two-stage H_2/CO-rich syngas production from endocarp coconut char (ECC) as the feedstock in continuous CO_2/O_2 gasification process. The gasification experiment was conducted at temperatures of 700 and 800 °C with a hold time of 30 min. The gasification agent in this experiment is a mixed gas (79/21%) of CO_2/O_2. The results showed that the yield of char and syngas was 60.2–72.6% and 27.4–39.8%, respectively, while the gas volume was 0.040–0.062 Nm^3. The average syngas composition was around 9.8–26.8% CO_2, 20.2–22.9% H_2, 42.9–58.2% CO, and 3.5–5.6% CH_4 depending upon the gasification temperature. The syngas composition ratios were 0.75–2.34 of H_2/CO_2 and 1.6–5.94 for CO/CO_2. Syngas lower heating value was around 9.6–11.1 MJ/Nm^3. Based on these findings, the continuous feed CO_2/O_2 biochar gasification could be used to produce H_2/CO-rich syngas for further conversion process. This underscores its viability for further technological development and optimization.

Keywords Gasification · Syngas · Biochar

3.1 Introduction

In recent years, demands for clean and renewable energy are growing to replace traditional fuels such as coal or oil that produce high emissions. Many renewable energy sources have been used to lessen energy demand, such as solar and wind. Beside those, many have shown growing interest in hydrogen production for renewable energy, which used in an energy production system such as fuel cells, after

N. Yutthakarn · C. Khaobang · C. Areeprasert (✉)
Department of Mechanical Engineering, Faculty of Engineering, Kasetsart University, Bangkok 10900, Thailand
e-mail: fengcta@ku.ac.th

S. Boonyubol · J. S. Cross
Department of Transdisciplinary Science and Engineering, School of Environment and Society, Institute of Science Tokyo, Tokyo 152-8550, Japan

C. Areeprasert et al. (eds.), *Proceedings of the International Conference on Sustainable Energy and Environmental Technology for Circular Economy*, Springer Proceedings in Energy, https://doi.org/10.1007/978-981-96-2306-8_3

the process, waste produced from this system are water vapor that is easier to treat than waste from traditional fuel. In addition, the thermal conversion process such as gasification could also be a renewable way for energy production using agricultural waste.

In 2020, Thailand produced approximately 817.8 million kilograms of coconut [1] which after process left the byproduct of endocarp coconut (EC) that EC can be used to produce biochar. From the study of Jigneshkumar Makwana [2], it was found that gasification is the best-suited technology to deal with combustible waste and turn them into useful syngas. Gasification is the process of turning carbonaceous materials into syngas under an oxygen-limited environment at high temperature. The syngas produced contains gases such as carbon monoxide, hydrogen, methane, and carbon dioxide. Other than that, the study of Anwar Sattar [3] showed the versatility of the gasification method for syngas production with biochar as fuel, which can produce high-quality syngas but is varied in the composition when using different biochar. Therefore, the use of biochar as fuel for gasification is a prominant choice for experiment due to wide variables of temperature, fuel type, and gasification agent to produce H_2/CO-rich syngas. With H_2/CO-rich syngas from gasification process, syngas upgrading could be done including water-gas shift process for hydrogen production. The study of Stan T. Kolaczkowski [4] showed that hydrogen composition in syngas can be separated and purified into pure hydrogen (>95 vol%). Therefore, the process of turning waste into uselful syngas is indeed possible.

In this study, the investigation of the gasification process with ECC as fuel with the use of the gasifying agent (79/21%) CO_2/O_2 has been conducted to study the feasibility for H_2/CO-rich syngas production to understand the process, composition of syngas, product yields, syngas heating value, and other parameters such as H_2/CO_2 and CO/CO_2 ratios. The main goal is to explore the process of H_2/CO-rich syngas production for future development of syngas upgrading process.

3.2 Material and Methods

3.2.1 Raw Material

Fuel used in this experiment is ECC obtained from a local farmer in Chachoengsao province, Thailand. The size of ECC was reduced by crushing into various sizes ranging from 12 to 15 mm with total weight of 200 g to be used in gasification process. After that, ECC was dried in oven at a temperature of 75 °C for 24 h. The properties of ECC are presented in Table 3.1.

Table 3.1 Properties of endocarp coconut char (ECC)

Item	ECC
Proximate analysis (wt.%)	
Moisture	7.8
Volatile material	22.2
Fixed carbon	67.4
Ash	2.6
Ultimate analysis (wt.%)	
Carbon	93.5
Hydrogen	2.01
Nitrogen	0.38
Oxygen	4.11
Heating value (MJ/kg)	
Higher heating value	26.14

3.2.2 *Gasification Process*

The gasification process in this experiment was conducted using a downdraft drop-tube gasifier set within an electric furnace, with temperatures set at 700 and 800°C for each experiment. Prior to initiating the experiment, the gasification system was purged of oxygen by introducing nitrogen at a controlled flow rate of 1 L/min. The purpose is to replace air in the gasifier reactor to prevent complete combustion. Oxygen levels were monitored using a TESTO 350 analyzer to confirm the absence of oxygen. Once the furnace reached the desired temperature, nitrogen flow was cut and replaced with a gasifying agent mixture of CO_2/O_2 (79/21%), maintaining the same flow rate. ECC was fed into the gasifier using an automatic feeding system, with a feed rate of 15 g/min and a total of 200 g of ECC for the experiment. The gasification process was held for 30 min, with syngas samples collected every 5-min intervals for gas composition analysis, and gas volume analysis conducted at 5, 15, and 30 min.

3.2.3 *Material Analysis Method*

Proximate analysis (ASTM D7582) was conducted using a NETZSCH STA 449 F3 Jupiter thermal analyzer. Elemental analysis (ASTM D5373) was conducted using the LECO 628 Series. The gas composition was analyzed using a micro-gas chromatograph (Agilent 490 Micro GC). The heating value of fuel was analyzed using the bomb calorimeter model Leco AC-500.

Table 3.2 Char and syngas product yields

Product	700 °C	800 °C
Char product (wt.%)	72.6	60.2
Syngas product (wt.%)	27.4	39.8
Syngas volume (Nm^3)	0.040	0.062

3.2.4 *Syngas Analysis*

The analysis of gas composition was done by Agilent Micro Gas Chromatography, which analyzed each gas into percentage of carbon dioxide, methane, oxygen, nitrogen, ethane and carbon monoxide. The percentage of syngas composition was used for calculation of Lower Heating Value (LHV) as shown in Eq. (3.1).

$$LHV = ((\%CO \times 126.36) + (\%H_2 \times 107.98) + (\%CH_4 \times 358.18)) \div 1000 \quad (3.1)$$

where $\%CO$, $\%H_2$, and $\%CH_4$ are the molar percentage of gases in syngas.

3.3 Result and Discussion

3.3.1 *Product Yield*

The product yield is shown in Table 3.2. It should be noted that tar product was negligible because biochar with low volatile matter has been used and it produced insignificant amount of tar. The maximum char and syngas products were 72.6% and 39.8% at 700 and 800 °C, respectively. The rapid decomposition of biomass charcoal accelerates pyrolysis, producing low molecular weight gases like carbon monoxide, hydrogen, and methane, leading to increasing biomass gas production [5]. At higher temperatures, the Boudouard reaction becomes more efficient, converting more carbon to gas and reducing the carbon content as it reacts with O_2 forming CO_2 [6]. Furthermore, gas volume measurements indicate an upward trend with increasing temperature.

3.3.2 *Syngas Composition*

Syngas composition and average syngas composition are presented in Fig. 3.1 and Table 3.3, respectively. The trend of CO and CO_2 shows the opposite results, as the experiment time increased, CO concentration increased while CO_2 concentration reduced. The maximum hydrogen concentration was achieved at the 10-min mark of the experiment. After that, declining trend of hydrogen contents was observed. Variations in temperature had a minimal impact on the gas composition's formation potential, though gasification at increasing temperature led to an increased yield of carbon monoxide. The observed increase in carbon monoxide can be attributed to the Boudouard reaction, where carbon in the biomass reacts with atmospheric carbon dioxide to produce carbon monoxide [7]. Conversely, fluctuations in hydrogen concentration may be linked to the tar-cracking process, wherein tar breaks down into carbon and hydrogen. Over time, as the available tar diminishes, the production of hydrogen gas decreases accordingly [8].

3.3.3 *Syngas Analysis*

The syngas composition ratios are presented in Table 3.4 for detailed analysis. The key composition ratios include H_2/CO_2 and CO/CO_2. The observed rise in the CO/CO_2 and H_2/CO_2 ratios signifies a shift in the underlying reaction mechanisms within the gasification process. The increase in carbon monoxide is likely attributed to the enhanced Boudouard reaction, which concurrently reduces carbon dioxide levels. Additionally, the heating value exhibits an upward trend with increasing temperature, as elevated gasification temperatures promote greater carbon monoxide production, thereby enhancing the heating value of the syngas.

3.4 Conclusions

This study investigated the production of H_2/CO-rich syngas through gasification using ECC as feedstock. Key findings from the experiments are as follows:

1. The decomposition efficiency of ECC improved at higher temperatures. At 800 °C, the process resulted in 39.8% syngas and 60.2% residual char in the gasifier. In contrast, at 700 °C syngas production was reduced to 27.4% with 72.6% of the material remaining as char.
2. Average hydrogen content in the syngas showed minimal variation at experiment temperatures of 700 °C and 800 °C, with hydrogen composition between 20.2% and 22.9% respectively.
3. The notable differences lied in the concentrations of carbon monoxide (CO) and carbon dioxide (CO_2). Lower temperatures favored CO_2 production (ranging from 9.8 to 26.8%), whereas higher temperatures enhanced CO formation, with concentrations ranging from 42.9 to 58.2%.
4. The H_2/CO_2 and CO/CO_2 ratios (ranging from 0.75–2.43 and 1.60–5.94, respectively) indicated that the Boudouard reaction occurred, reducing CO_2 levels and increasing CO content. The LHV of syngas was around 9.6 to 11.1 MJ/Nm^3.

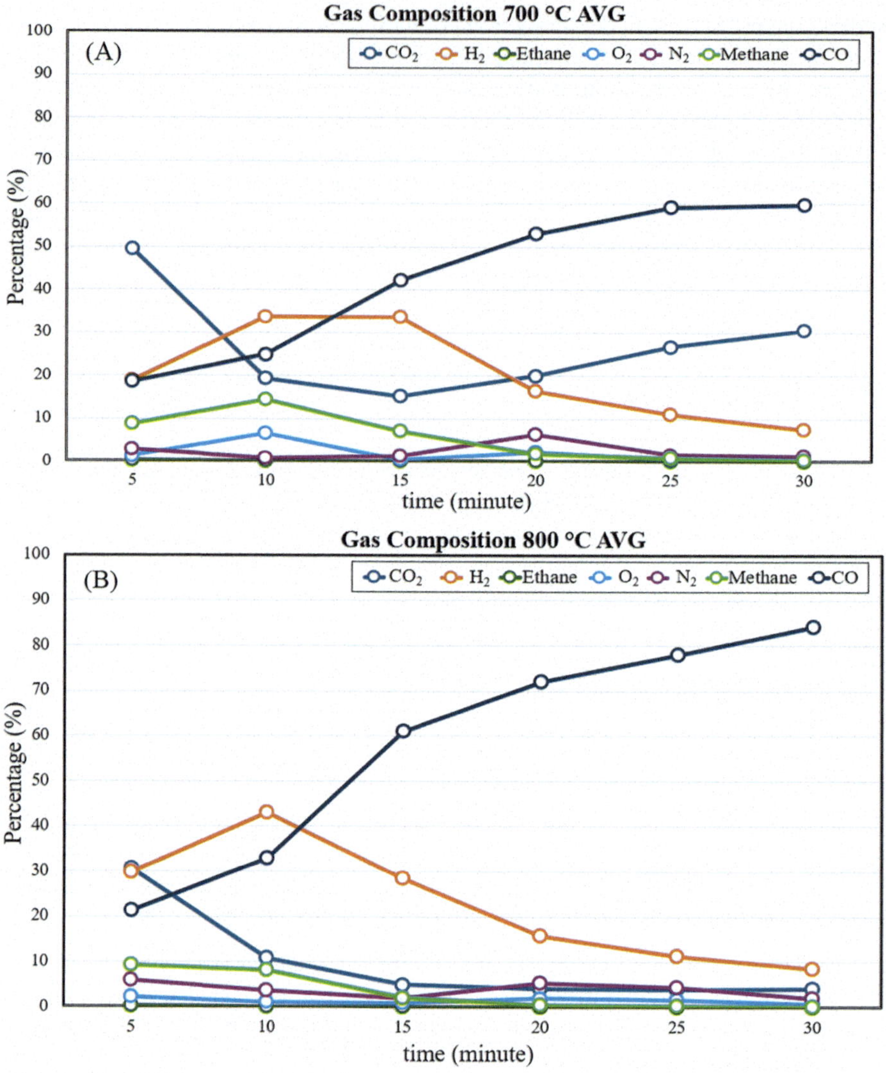

Fig. 3.1 Gas composition from gasification with ECC as fuel at **a** 700 °C and **b** 800 °C

Table 3.3 Average syngas composition at different gasification temperatures during data collection (vol%)

Average syngas composition	700 °C	800 °C
Carbon dioxide (CO_2)	26.8	9.8
Hydrogen (H_2)	20.2	22.9
Carbon monoxide (CO)	42.9	58.2
Methane (CH_4)	5.6	3.5

Table 3.4 Syngas composition ratios and its heating value

Item	700 °C	800 °C
CO/CO_2	1.60	5.94
H_2/CO_2	0.75	2.34
LHV (MJ/Nm^3)	9.6	11.1

Thus, the gasification of ECC in CO_2/O_2 medium proved to be working for H_2/CO-rich syngas production which could provide the reliable syngas source for further syngas upgrading in the future study.

Acknowledgements This work received funding from the Kasetsart University Research and Development Institute (KURDI, FF(KU) 51.67). In addition, this work was supported in part by the World Research Hub (WRH) Program of the International Research Frontiers Initiative, Institute of Science Tokyo. This work also received funding from Kasetsart University Postdoctoral Fellowship Program 2024. N. Yutthakarn participated in the Young Science and Engineering Researchers Program (YSEP) under the collaboration between the Institute of Science Tokyo and Kasetsart University and received scholarship from JASSO.

References

1. Department of Agriculture of Thailand, Coconut production situation (in Thai) (2020)
2. J. Makwana, A.D. Dhass, P.V. Ramana, D. Sapariya, D. Patel, An analysis of waste/biomass gasification producing hydrogen-rich syngas: a review. Int. J. Thermofluids (20), 100492 (2023)
3. A. Sattar, G.A. Leeke, A. Hornung, J. Wood, Steam gasification of rapeseed, wood, sewage sludge and miscanthus biochars for the production of a hydrogen-rich syngas. Biomass Bioenerg. **69**, 276–286 (2014)
4. S.T. Kolaczkowski, P. Willacy, M.D. Jones, M. Pugh, Separating hydrogen from syngas (produced from the gasification of waste) using pressure swing adsorption. Chem. Eng. Trans. **105**, 91–96 (2023)
5. N.S. Muhammed, A.O. Gbadamosi, E.I. Epelle, A.A. Abdulrasheed, B. Haq, S. Patil, M.S. Kamal, Hydrogen production, transportation, utilization, and storage: recent advances towards sustainable energy. J. Energy Storage **73**, 109207 (2023)
6. P.R. Lanjekar, N.L. Panwar, C. Agrawal, A comprehensive review on hydrogen production through thermochemical conversion of biomass for energy security. Bioresour. Technol. Rep. **21**, 101293 (2023)
7. M. Sieradzka, A. Mlonka-Mędrala, A. Magdziarz, Comprehensive investigation of the CO_2 gasification process of biomass wastes using TG-MS and lab-scale experimental research. Fuel **330**, 125566 (2022)
8. C. Deng, W. Song, Z. Chai, S. Guo, Z. Zhu, Characteristics of tar thermal cracking and catalytic conversion during circulating fluidized bed char gasification. Energy Fuels **34**(1), 142–149 (2022)

Chapter 4
Co-Pyrolysis of Waste Plastics and Used Engine Oil to Produce Alternative Liquid Fuel

Mochamad Syamsiro, Muhammad Arief Saputro, Bayu Megaprastio, Nanda Agusta Virgiawan, and Zahrul Mufrodi

Abstract The objective of this study is to examine the thermal co-pyrolysis process of waste plastics and used engine oil (UEO) as the effect of plastic material, temperature, and reactor diameter. High density polyethylene (HDPE), polypropylene (PP), and polystyrene (PS) have been used as waste plastic samples. A 400-g sample, consisting of 250 g of waste plastic and 150 g of used engine oil, was introduced into the vertical tubular reactor. The experimental results show that the liquid yield produced during HDPE/UEO co-pyrolysis at a temperature of 450 °C is the lowest. At the same temperature, PP/UEO and PS/UEO co-pyrolysis generate a greater amount of liquid compared to HDPE/UEO. The results of the co-pyrolysis of waste plastic and UEO were subsequently compared to the results of the individual pyrolysis of waste plastic and UEO. The result shows that more liquid was produced by co-pyrolysis for all types of plastic at temperatures of 450 °C and 550 °C than by individual pyrolysis. The experiment's results also show that co-pyrolyzing waste plastics and UEO produces more liquid at smaller diameters. The highest liquid yield of 69.85, 88.53, and 87.53% was obtained from the co-pyrolysis of HDPE/UEO, PP/UEO, and PS/UEO at a reactor diameter of 60 mm. In the case of co-pyrolysis of HDPE/UEO, the production of liquid increases with decreasing reactor diameter, increasing by 9.8% from 60 to 69.8%. Higher results were obtained from co-pyrolysis of PP/UEO and PS/UEO, with an increase of 27% for PP/UEO and 31.8% for PS/UEO.

Keywords Co-pyrolysis · Waste plastics · Used engine oil · Temperature · Reactor diameter

M. Syamsiro (✉) · M. A. Saputro · B. Megaprastio · N. A. Virgiawan
Department of Mechanical Engineering, Janabadra University, Yogyakarta 55231, Indonesia
e-mail: syamsiro@janabadra.ac.id

M. Syamsiro · M. A. Saputro · B. Megaprastio
Center for Waste Management and Bioenergy, Janabadra University, Yogyakarta 55231, Indonesia

Z. Mufrodi
Department of Chemical Engineering, Universitas Ahmad Dahlan, Yogyakarta 55191, Indonesia

C. Areeprasert et al. (eds.), *Proceedings of the International Conference on Sustainable Energy and Environmental Technology for Circular Economy*, Springer Proceedings in Energy, https://doi.org/10.1007/978-981-96-2306-8_4

4.1 Introduction

Solid waste management is a critical issue related to the economic development and lifestyle of a population. As developing nations, especially those in the ASEAN region, experience rapid industrialization and urbanization, they are grappling with the immense task of effectively managing the ever-increasing volumes of municipal solid waste (MSW) and plastic waste [1, 2]. Indonesia is considered the world's second-largest plastic polluter, with an estimated 3.2 million tonnes of plastic waste entering the ocean annually [3]. Plastic waste management in Indonesia and other ASEAN countries faces significant challenges due to the lack of adequate waste collection, recycling infrastructure, and public awareness [4]. Therefore, understanding the current state of solid waste and plastic waste management, as well as exploring sustainable solutions, is crucial for addressing this pressing environmental issue in developing nations.

Simultaneously, Indonesia's economic progress in recent years has enhanced the movement of individuals and commodities. Therefore, there is a growing demand for additional transportation options, particularly for land-based modes such as railways and roads [5]. This includes the need for passenger vehicles, trucks, buses, and motorcycles. According to the data provided by Badan Pusat Statistik (BPS), the average rate of increase in motor vehicles between 2018 and 2022 exceeded 4%, with motorcycles experiencing the highest growth rate of nearly 4.4% [6]. Consequently, the transportation modes mentioned will produce a greater amount of used engine oil.

On the other hand, the increasing demand for energy and the growing environmental concerns surrounding plastic waste disposal have prompted researchers to explore alternative solutions for waste management and energy production. One promising approach is the co-pyrolysis of waste plastics and used engine oil to produce alternative liquid fuels [7, 8]. The conversion of plastic waste into energy through pyrolysis has been extensively studied, as plastics are derived from petroleum and the resulting oil has a high calorific value [9–12]. Utilization of the liquid fuel from waste plastic in diesel engine has also been investigated by some researcher [13, 14]. In addition to plastic waste, used engine oil has also garnered attention as a potential feedstock for fuel production due to its hydrocarbon content, which can be leveraged to create alternative liquid fuels through co-pyrolysis processes.

By utilizing co-pyrolysis to convert these waste streams into alternative liquid fuels, ASEAN countries can address both waste management and energy security concerns. The development of co-pyrolysis technologies for the production of alternative liquid fuels from waste plastics and used engine oil in ASEAN countries holds great potential and warrants further investigation and investment to support the region's transition towards a more environmentally responsible and energy-efficient future. Therefore, the current study investigated the thermal co-pyrolysis of waste plastics and used engine oil to produce alternative liquid fuel which can be used as a partial substitute for gasoline or diesel fuel.

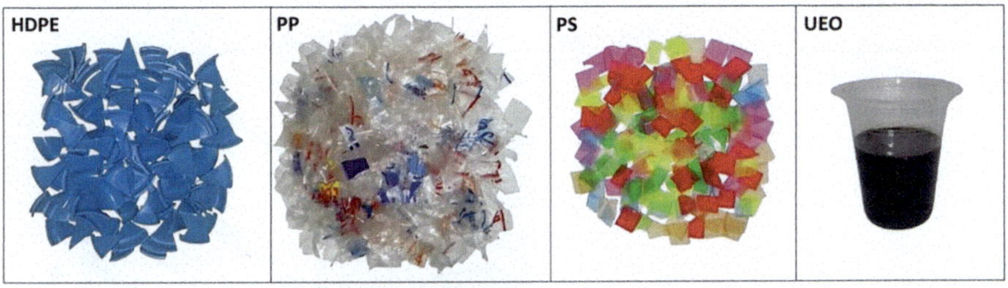

Fig. 4.1 Crushed waste plastic materials (HDPE, PP, and PS) and used engine oil (UEO) as feedstocks for co-pyrolysis study

4.2 Materials and Methods

4.2.1 Materials

The waste plastics used in this investigation included high density polyethylene (HDPE), polypropylene (PP), and polystyrene (PS) collected from a local recycling company. Used engine oil (UEO) was obtained from a local automobile repair shop. The plastic materials were subsequently crushed into smaller size to maximize the capacity of the reactor. Figure 4.1 shows the appearance of the feedstocks.

4.2.2 Experimental Procedure

The pyrolysis experiments were carried out on a small scale vertical tubular reactor made from stainless steel. The reactor was heated with an electrical heater. The height of the pyrolysis reactor is 450 mm. The reaction temperatures in the reactor were measured using a K-type thermocouple and controlled by the heater. The condenser was attached to the reactor's exit to condense the pyrolysis gas. The first condenser was cooled by the ambient air, while the second is a double-pipe condenser that was cooled by the circulating water. Figure 4.2 shows a three-dimensional schematic diagram of the equipment. The liquid product was then collected at the bottom of the oil tank's condenser. The solid residue remained in the pyrolyzer and was weighed to determine the mass balance.

A 400-g sample, consisting of 250 g of waste plastic and 150 g of used engine oil, was introduced into the reactor simultaneously in each experiment. The reactor was thereafter closed and heated from ambient temperature to the desired temperature. The waste plastic was subsequently melted and volatilized in conjunction with UEO, resulting in a gaseous phase. This gas was then directed to the condenser for cooling and condensing, resulting in the formation of pyrolytic oil. The thermal pyrolysis studies were conducted at reactor temperatures of 450, 500, and 550 °C. Experiments were also conducted for three different reactor diameters: 60, 80, and 100 mm.

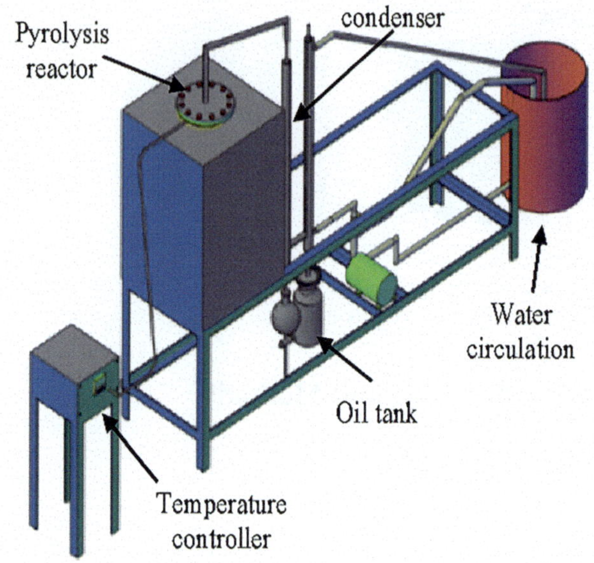

Fig. 4.2 A schematic diagram of the pyrolysis experimental apparatus [9]

4.3 Results and Discussions

4.3.1 Effect of Plastic Types

The effect of pyrolysis temperature on liquid yield as a function of time can be seen in Fig. 4.3. A pyrolysis reactor with a diameter of 100 mm was used in this study. The experimental results show that a temperature of 450 °C takes the longest time for the pyrolysis process. The higher the reactor temperature, the faster the pyrolysis process will take place. The liquid yield produced during HDPE/UEO co-pyrolysis at a temperature of 450 °C is the lowest. This is due to the insufficient energy generated at that temperature to activate HDPE plastic. This is evident from the significant amount of solid residue, indicating that not all of the HDPE plastic has undergone pyrolysis. At the same temperature, PP/UEO and PS/UEO co-pyrolysis generate a greater amount of liquid compared to HDPE/UEO. This is due to the fact that the activation energy for PP and PS is lower than that of HDPE.

The maximum liquid yield was achieved at a temperature of 500 °C in HDPE/UEO co-pyrolysis, while the highest liquid yield was at 450 °C in PP/UEO and PS/UEO co-pyrolysis. The lowest yield was obtained at a temperature of 550 °C for PP/UEO and PS/UEO. This indicates that at this temperature, a larger proportion of the pyrolysis products exist in the form of gas due to the higher temperature causing the hydrocarbon chain to be cracked into shorter segments.

The results of the co-pyrolysis of waste plastic and UEO were subsequently compared to the results of the individual pyrolysis of waste plastic and UEO to observe the synergistic effect of those materials. The results of the comparison of

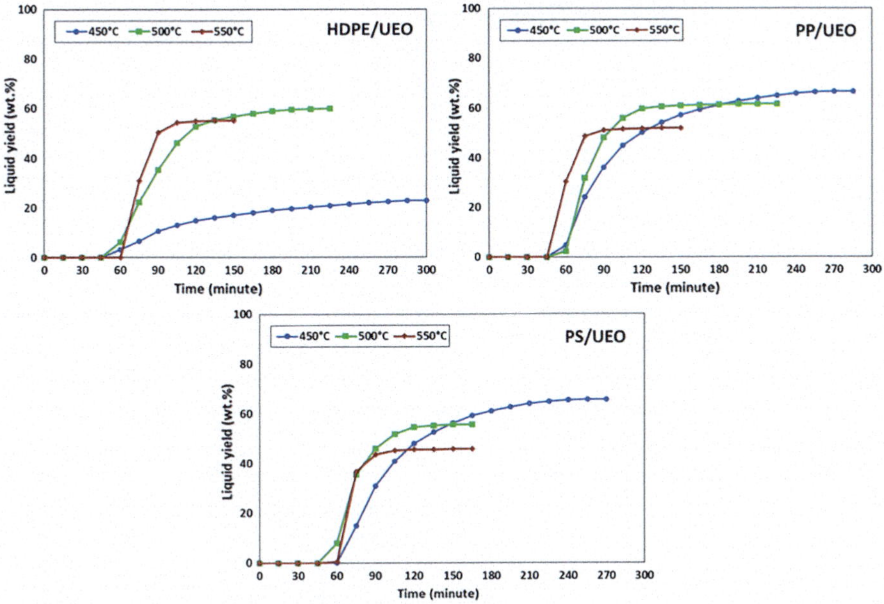

Fig. 4.3 Effect of reactor temperature on liquid yield from co-pyrolysis of waste plastics (HDPE, PP, PS) and used engine oil (UEO) in a fixed bed reactor

co-pyrolysis from experiments and calculations are illustrated in Fig. 4.4. The figure shows that more liquid was produced by co-pyrolysis for all types of plastic at temperatures of 450 and 550 °C than by individual pyrolysis. Different results were obtained at a temperature of 500 °C where co-pyrolysis produced lower liquid.

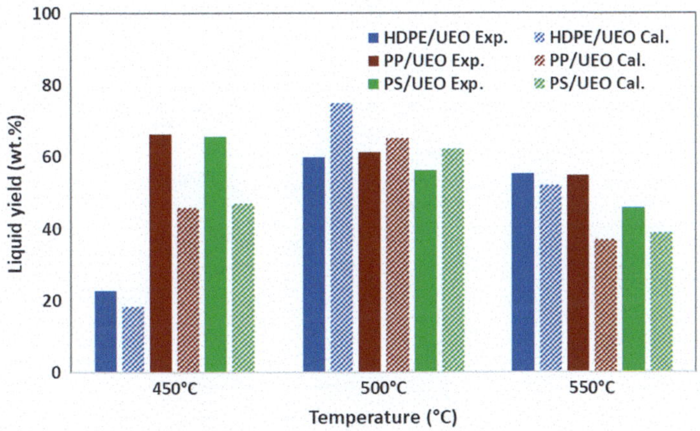

Fig. 4.4 Comparison of co-pyrolysis of waste plastic and UEO (experimental-Exp.) and calculation from individual pyrolysis of waste plastic and UEO (calculation-Cal.)

4.3.2 Effect of Reactor Dimension

The effect of reactor diameter on liquid yield as a function of time is depicted in Fig. 4.5. The experiment's results show that co-pyrolyzing waste plastics and UEO produces more liquid at smaller diameters. Additionally, the smaller diameter suggests a quicker pyrolysis time. This is due to the wider contact surface of the feedstock with the smaller diameter, which means that the heat transfer process into the reactor becomes higher. As a result, more energy enters the plastic melting process, as well as the evaporation of plastic and UEO.

The highest liquid yield of 69.85%, 88.53%, and 87.53% was obtained from the co-pyrolysis of HDPE/UEO, PP/UEO, and PS/UEO at a reactor diameter of 60 mm. This is crucial to consider while designing a pyrolysis reactor where the con-tact area of the feedstock with the reactor is one of the keys in the pyrolysis process to obtain maximum liquid yield.

Figure 4.6 depicts a comparison of liquid yield for co-pyrolysis of HDPE/UEO, PP/UEO, and PS/UEO due to the influence of reactor diameter size. The picture illustrates that in the case of co-pyrolysis of HDPE/UEO, the production of liquid increases with decreasing reactor diameter, increasing by 9.8% from 60 to 69.8%. Higher results were obtained from co-pyrolysis of PP/UEO and PS/UEO, with an increase of 27% for PP/UEO and 31.8% for PS/UEO.

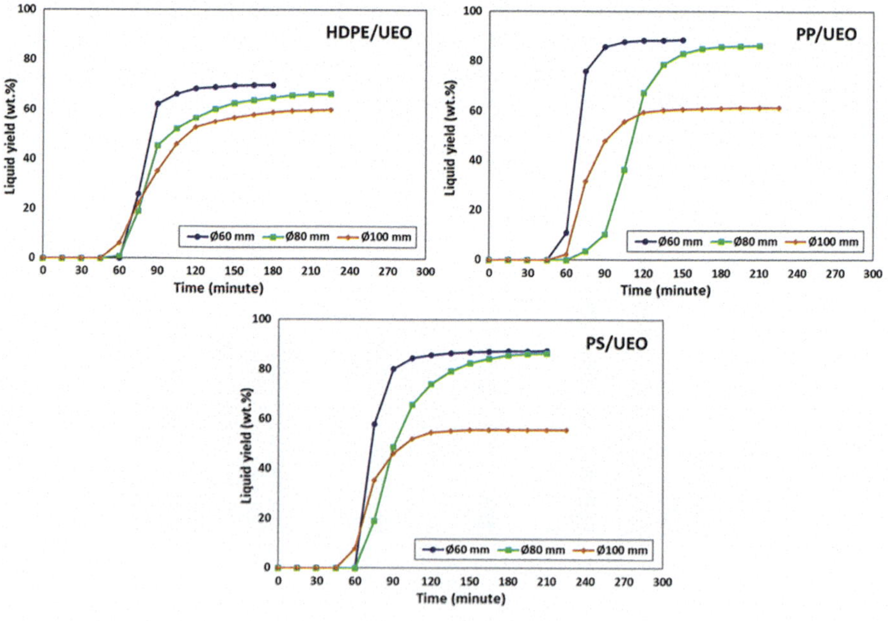

Fig. 4.5 Effect of reactor diameter on liquid yield from co-pyrolysis of waste plastics (HDPE, PP, PS) and used engine oil in a fixed bed reactor

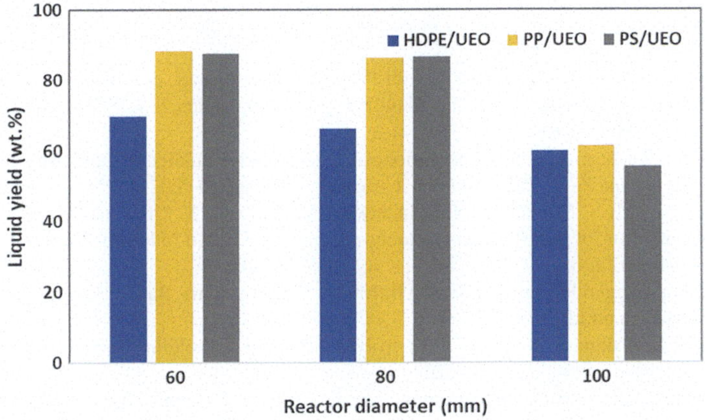

Fig. 4.6 Comparison of liquid yield from co-pyrolysis of HDPE/UEO, PP/UEO, and PS/UEO as the effect of reactor diameter

4.4 Conclusion

Co-pyrolysis of waste plastics and UEO has been investigated as the effect of reactor temperature and diameter. The experimental results show that a temperature of 450 °C takes the longest time for the pyrolysis process. The liquid yield produced during HDPE/UEO co-pyrolysis at a temperature of 450 °C is the lowest. At the same temperature, PP/UEO and PS/UEO co-pyrolysis generate a greater amount of liquid compared to HDPE/UEO. The results of the co-pyrolysis of waste plastic and UEO were subsequently compared to the results of the individual pyrolysis of waste plastic and UEO to observe the synergistic effect of those materials. The result shows that more liquid was produced by co-pyrolysis for all types of plastic at temperatures of 450 and 550 °C than by individual pyrolysis. Different results were obtained at a temperature of 500 °C where co-pyrolysis produced lower liquid. The experiment's results also show that co-pyrolyzing waste plastics and UEO produces more liquid at smaller diameters. The highest liquid yield of 69.85, 88.53, and 87.53% was obtained from the co-pyrolysis of HDPE/UEO, PP/UEO, and PS/UEO at a reactor diameter of 60 mm. In the case of co-pyrolysis of HDPE/UEO, the production of liquid increases with decreasing reactor diameter, increasing by 9.8% from 60 to 69.8%. Higher results were obtained from co-pyrolysis of PP/UEO and PS/UEO, with an increase of 27% for PP/UEO and 31.8% for PS/UEO.

References

1. H.L. Chen, T.P. Nath, S. Chong, V. Foo, C. Gibbins, A.M. Lechner, The plastic waste problem in Malaysia: management, recycling and disposal of local and global plastic waste. SN Appl. Sci. **3**, 1–15 (2021)

2. R. Chandrappa, D. Das, *Solid waste management* (Springer, Berlin, Heidelberg, 2012). https://doi.org/10.1007/978-3-642-28681-0
3. W. Setyaningsih, H. Hadiyanto, T. Triadi Putranto, Microplastic pollution in Indonesia: The contribution of human activity to the abundance of microplastics, in E3S Web of Conferences, vol. 448 (2023) pp. 1–10
4. A, Aprilia, Waste management in Indonesia and Jakarta: challenges and way forward, in *Proceedings of the 23rd ASEF Summer University, Virtual, 20* (2021)
5. M. Syamsiro, M. S. Dwicahyo, Y. Sulistiawati, M. Ridwan, N. Citrasari, Development of a rotary kiln reactor for pyrolytic oil production from waste tire in Indonesia, in *IOP Conference Series: Earth and Environmental Science* vol. 245 (2019)
6. BPS, Perkembangan Jumlah Kendaraan Bermotor Menurut Jenis. Badan Pusat Statistik (2023), https://www.bps.go.id/
7. S. Uçar, A.R. Özkan, S. Karagöz, Co-pyrolysis of waste polyolefins with waste motor oil. J. Anal. Appl. Pyrol. **119**, 233–241 (2016)
8. W. Charusiri, N. Phowan, A. Permpoonwiwat, T. Vitidsant, Catalytic copyrolysis of used waste plastic and lubricating oil using cu-modification of a spent fluid catalytic cracking catalyst for diesel-like fuel production. ACS Omega **8**, 40785–40800 (2023)
9. M. Syamsiro, Z. Mufrodi, R. Rafly, S. Machmud, Energy recovery from food packaging plastics by thermal and catalytic pyrolysis processes. Univers. J. Mech. Eng. **8**, 51–58 (2020)
10. M. Syamsiro, D.Y. Mathias, H. Saptoadi, D.R. Sawitri, A.S. Nizami, M. Rehan, Pyrolysis of compact disc (CD) case wastes to produce liquid fuel as a renewable source of electricity generation. Energy Procedia **145**, 484–489 (2018)
11. M.H. Rahman, P.R. Bhoi, P.L. Menezes, Pyrolysis of waste plastics into fuels and chemicals: a review. Renew. Sustain. Energy Rev. **188** (2023)
12. O.Y. Yansaneh, S.H. Zein, Recent advances on waste plastic thermal pyrolysis: a critical overview. Processes **10**(332), 1–25 (2022)
13. M. Syamsiro, H. Saptoadi, M. Kismurtono, Z. Mufrodi, K. Yoshikawa, Utilization of waste polyethylene pyrolysis oil as partial substitute for diesel fuel in a DI diesel engine. Int. J. Smart Grid Clean Energy **8**(1), 38–47 (2019)
14. S. Kanchan, M. Priyadarshini, P. Kumar, R. Choudhary, S. Pradhan, Multi-objective statistical optimisation utilising response surface methodology to predict engine performance using biofuels from waste plastic oil in CRDi engines. Green Process. Synth. **13** (2024)

Part II
Material from Biomass and Waste

Chapter 5
Surface-Modified Synthesis of Low-Cost Adsorbents Derived from Fly Ash for Removing Ammonia

Tulakarn Ketwong, Natchapon Cholwatthanatanakorn, and Chinnathan Areeprasert

Abstract Ammonia (NH_3) emissions from various agricultural and industrial activities pose significant environmental and health concerns, necessitating effective removal methods. This study focuses on utilizing waste materials, specifically fly ash (FA) derived from sugarcane bagasse incineration, combined with bamboo hydrochar (BH) or red mud (RM), for NH_3 adsorption. The FA was physically activated to enhance its porosity and surface area, while BH and RM were incorporated to improve NH_3 adsorption capabilities. The adsorbents were characterized using SEM and BET surface analysis. NH_3 adsorption experiments were conducted, revealing that AFA-BH exhibited the highest adsorption capacity at 0.84 mL-NH_3/g-adsorbent, attributed to chemical adsorption facilitated by functional groups in BH. AFA-RM also demonstrated significant NH_3 adsorption at 0.75 mL-NH_3/g-adsorbent due to the presence of iron oxide in RM. Steam activation increased micropore volume and specific surface area, enhancing NH_3 adsorption of FA. This research offers a low-cost and environmentally friendly approach to NH_3 removal, utilizing waste materials and avoiding chemical modifications.

Keywords Ammonia adsorption · Adsorbent · Fly ash · Activation

T. Ketwong · N. Cholwatthanatanakorn · C. Areeprasert (✉)
Department of Mechanical Engineering, Faculty of Engineering, Kasetsart University, 50 Ngam Wong Wan Road, Lat Yao, Chatuchak, Bangkok 10900, Thailand
e-mail: fengcta@ku.ac.th

C. Areeprasert et al. (eds.), *Proceedings of the International Conference on Sustainable Energy and Environmental Technology for Circular Economy*, Springer Proceedings in Energy, https://doi.org/10.1007/978-981-96-2306-8_5

5.1 Introduction

NH_3 has emerged as a crucial natural element across various agricultural and industrial sectors, including fertilizer production, pharmaceuticals, antifreeze, heat pumps, and proton transportation. Its widespread use results in its dispersion into the atmosphere, leading to indoor air pollution as it slowly releases from surfaces due to environmental factors like temperature and humidity. Additionally, ammonia contributes to the formation of aerosols and fine particles, playing an important role in the creation of PM2.5 pollution [1]. The pungent odor of ammonia poses health risks, damaging both the skin and respiratory systems and disrupting ecological balance and biodiversity. Environmental ammonia levels typically range from 0.5 to 1.0 mg/m^3, with exposure to concentrations exceeding 30 mg/m^3 for over 15 min resulting in severe lung cell damage [2]. Notably, ammonia is highly reactive, forming explosive gases when reaching air volumes of 16.5–26.5%. Consequently, many countries have implemented stringent regulations to control ammonia concentrations in ambient air, highlighting the need for effective removal of ammonia from indoor and industrial environments.

Many techniques are utilized for NH_3 removal, including biological filtration, adsorption, and catalysis. Among these, adsorption proves to be the most efficient, offering convenient operation and environmental friendliness. Numerous adsorbents had been produced from agricultural or industrial residue, offering numerous advantages due to their versatility in applications such as supercapacitors [3], and gas adsorption [4]. The effectiveness of these adsorbents depends significantly on their considerable porosity and surface area, and they could be manufactured from various biomass sources like bamboo and bagasse. Several researches had been conducted on utilizing agricultural and industrial residue for NH_3 removal. In 2021, Ren et al. fabricated an adsorbent enhanced with $Fe(NO_3)_3$ impregnation followed by calcination to capture NH_4^+ in wastewater. The findings demonstrated that the adsorbent's capacity increased from 0.088 to 0.15 mg/g after Fe loading [5]. Bai et al. (2021) had transformed rice husk into an adsorbent to eliminate airborne ammonia [6]. These investigations provide compelling evidence that specific waste materials could be transformed into efficient NH_3 adsorbents.

The utilization of FA derived from sugarcane bagasse incineration in biomass power plants presented a promising method for the creation of absorbent materials. In the period of 2021/2022, Thailand cultivated approximately 4.4 million acres of sugarcane, yielding an output of 105.9 million tons of sugarcane [7]. FA had undergone the necessary thermal and chemical transformation processes during burning, rendering it suitable for adsorbent production derived from biomass. FA could be straightly employed without additional thermal treatments, unlike raw biomass. FA needs only activation or modification processes to enhance its efficacy. This research not only investigates the production of NH_3 adsorbent derived from FA via steam activation but also explores methods for functionalizing them. The study also focuses on utilizing agricultural and industrial waste materials to enhance the absorption of NH_3. Bamboo and red mud (RM), which is a byproduct of alumina purification from bauxite, were considered as potential sources. RM is rich in iron oxide (Fe_2O_3), with about 65–70% content. It could significantly improve NH_3 absorption through chemisorption. Bamboo, which contributes to about 50% of waste during chopstick production in Thailand, was previously disposed of through incineration. However, it possesses many functional groups like hydroxyl (-OH), carbonyl (-C=O), and carboxyl (-COOH) that could chemically bond with molecules of NH_3. The content of carbon in bamboo could be enhanced through hydrothermal carbonization (HTC), a process converting biomass into a carbon-rich material called hydrochar, which developed NH_3 adsorption. Therefore, combining these materials with FA was beneficial since they could enhance the overall performance of the absorbent.

The objective of this study is to create NH_3 adsorbents from waste materials, specifically FA activated through steam activation followed by improvement with BH or RM, using cost-effective processes without adding any chemicals. Both surface-modified absorbents were analyzed and compared. The adsorbents' NH_3 adsorption capabilities were tested using an ammonia solution. This research offers potential waste management solutions by repurposing waste into NH_3 adsorbents applicable to the world's environment.

5.2 Methodology

5.2.1 Material

FA was provided from Mitr Phol Bio-Power (Phu Viang) Co., Ltd., located in Khon Kaen. RM was obtained from Sweet Mold and Soap Supplies Co., Ltd. Bamboo scraps were obtained from chopstick factory waste in Chonburi province. AR-grade ammonium hydroxide (NH_4OH) was purchased from Fisher Chemical, and a diluted NH_4OH solution was prepared using deionized water.

5.2.2 Fabrication of Adsorbent

In this research, the adsorbents were synthesized through two main processes: hydrothermal carbonization (HTC) of bamboo to produce bamboo hydrochar (BH) and subsequent activation of the precursors. Figure 5.1 depicts visual representations of both the preparation of material and the setup of NH_3 adsorption experiment.

Hydrothermal Carbonization

Initially, bamboo scrap was ground into a powdered form. Then, it was filtered to achieve a 100-mesh particle size. Subsequently, 5 g of the crushed bamboo were combined with 100 mL of deionized water. The combination was stirred employing

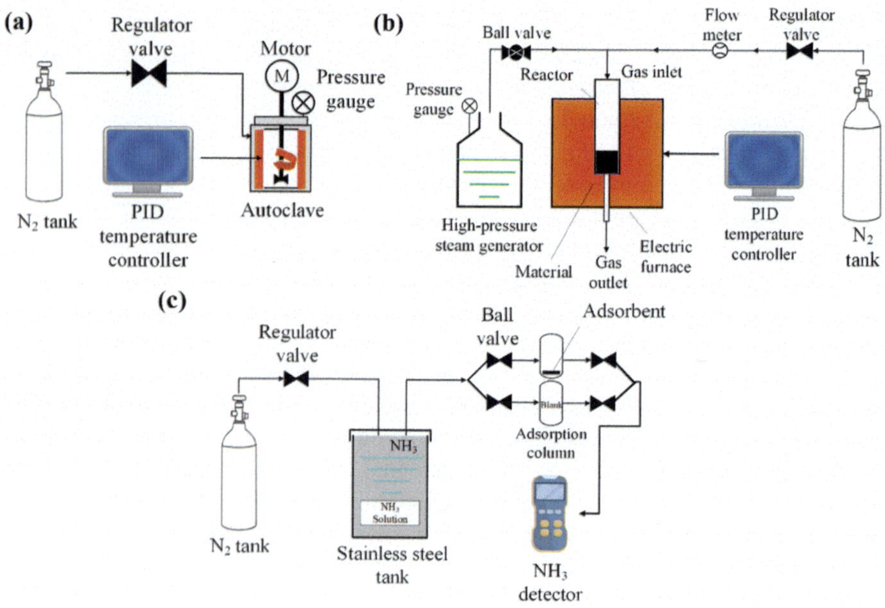

Fig. 5.1 **a** HTC procedure; **b** activation system; **c** equipment of NH_3 adsorption experiment

a magnetic stirrer and transferred to a 0.5-L HTC reactor. Before commencing the process, the reactor was fed with N_2 in order to eliminate O_2 that remained inside. Afterwards, the mixture was steadily heated up to 200 °C with a ramp rate of 10 °C/min and held for 1 h. After the completion of the experiment, the heating system was shut down, and the reactor was allowed to cool naturally at a rate of about 1 °C/min. Once the temperature dropped below 60 °C/min, the solid phase of the product was collected from the HTC reactor and oven-dried at 105 °C overnight. Figure 5.1a illustrates the HTC apparatus's diagram utilized for this investigation.

Steam Activation Process

The steam activation was conducted in order to enhance the material's porosity and surface area. Firstly, 10 g of FA was inserted into the vertical reactor illustrated in Fig. 5.1b. Steam was generated using an All-American electric sterilizer (Model 25X-2) as a steam generator, producing steam at temperatures of approximately 120 °C. Thereafter, the PID temperature controller was applied to heat the reactor from ambient temperature to the target temperature of 700 °C with a heating rate of 10 °C/min. This terminal temperature was held for a duration of 1 h. Throughout the activation process, N_2 was introduced into the reactor in order to avoid combustion. Once the activation was finished, the product was brought out and kept in a desiccator. The product obtained from steam activation was represented as AFA. For the combination FA with BH or RM, both surface modifiers were initially pulverized into powder and then combined with FA in a weight percentage of 85:15% (FA to

BH or RM) to serve as precursors for the activation process. Afterward, the combination underwent activation utilizing the same process employed for AFA fabrication. The adsorbents comprising of FA and BH or RM were designated as AFA-BH and AFA-RM, respectively.

5.2.3 *Material Characterization*

The morphological characteristics of the samples were examined using a JEOL scanning electron microscope (SEM) (Model JSM-6610LV). The porosity and textural studies of the material were investigated through Quantachrome Brunauer-Emmett-Teller (BET) surface analyzer. The amount of removed NH_3 of the fabricated adsorbents was evaluated using the equipment of the NH_3 adsorption experiment setup illustrated in Fig. 5.1c. 0.5 L/min-N_2 was utilized as the carrier for NH_3 sourced from a stainless steel tank (SUS 304). The flow of gas was monitored and regulated by applying a flow meter. 1.3 ml of 28% NH_4OH was added to 1 L of deionized water to prepare a dilute ammonia solution with a concentration of 0.04 mg/mL (~40 ppm). This dilute solution was added to the stainless steel tank, while the adsorption column was loaded with 2 g of the adsorbent. Firstly, NH_3 flowed into another column with no adsorbent, whereas the valve connected to the adsorbent column was still closed. The concentration of NH_3 was measured every 1 min for half an hour via a smart sensor NH_3 gas detector (AR8500). Another column without adsorbent was used for 20 min for confirmation of NH_3 concentration stability before the measurement. After that, the gas flow was redirected from the empty column to the column containing the adsorbent by closing the valve connected to the empty column and opening the valve of the adsorption column instead. At this stage, NH_3 flowed into the adsorption column. The area under each plot was used to compute the amount of adsorbed NH_3 in each adsorbent.

5.3 Results and Discussion

The SEM technique was applied to observe the surface structure and morphology of FA and the produced adsorbents. Figure 5.2a–d display the SEM images of the materials. In Fig. 5.2a, the morphology of FA reveals an irregular shape with sporadic pores on a mostly smooth surface. Figure 5.2b demonstrates the impact of activation, resulting in more pores of relatively uniform diameter compared to untreated FA, induced by the activation process. AFA-BH surface displayed bamboo fiber bundles (long rods), indicating thorough mixing with FA. The structure of AFA-RM was depicted in Fig. 5.2d, appearing as irregular shapes of RM scattered across the adsorbent surface, which was similar to the work conducted by Li et al. [8]. RM presence was expected to enrich the adsorbent with more Fe and Al content. These SEM findings validated the enhancement of the adsorbent derived from FA and its

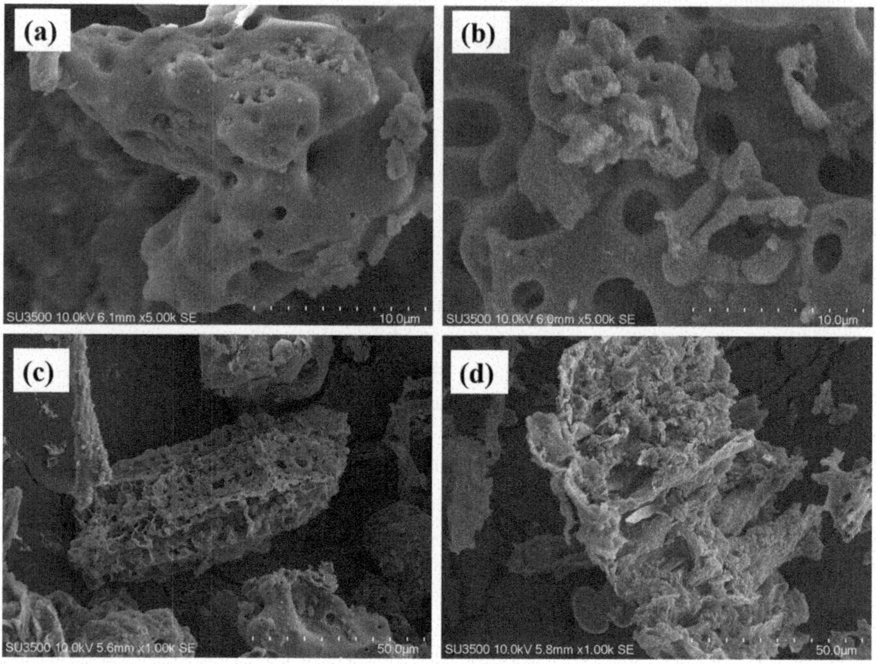

Fig. 5.2 SEM images of **a** FA; **b** AFA; **c** AFA-BH; **d** AFA-RM

successful blending with bamboo fiber particles and RM, potentially benefiting the NH_3 adsorption efficacy.

The results illustrated in Fig. 5.3a demonstrated a noteworthy trend in NH_3 removal rates over time, with a steady increase followed by a peak at the 5th min and a subsequent decline. To simplify the results, the area under the curves was computed to derive the NH_3 adsorption capacities, as depicted in Fig. 5.3b. The result indicated that AFA exhibited NH_3 adsorption with a capacity of 0.59 mL-NH_3 per g of adsorbent (g-adsorbent). Notably, AFA-BH displayed the highest NH_3 adsorption capacity, reaching 0.84 mL-NH_3/g-adsorbent after a half-hour adsorption experiment, which was attributed mostly to chemical adsorption distributed by the functional groups from BH (carboxyl, -COOH and hydroxyl, -OH). It was higher than that of AFA-RM, whose capacity was 0.75 mL-NH_3/g-adsorbent. In brief, these adsorbents demonstrated a multifaceted mechanism involving both chemical and physical adsorption. Physisorption primarily was from weak van der Waals forces, while chemisorption was largely attributed to the presence of functional groups such as -COOH and -OH, as well as ion exchange processes. Specifically, hydroxyl and carboxyl groups are capable of adsorbing NH_3, leading to the formation of NH_4^+ on Brønsted acid sites via H bonding, where the nitrogen atom is bonded to the H atom of -COOH or -OH [9].

BET specific surface area (S_{BET}) and total pore volume (V_{total}) displayed a noticeable decrease after the loading of BH or RM into FA, which was attributed to their

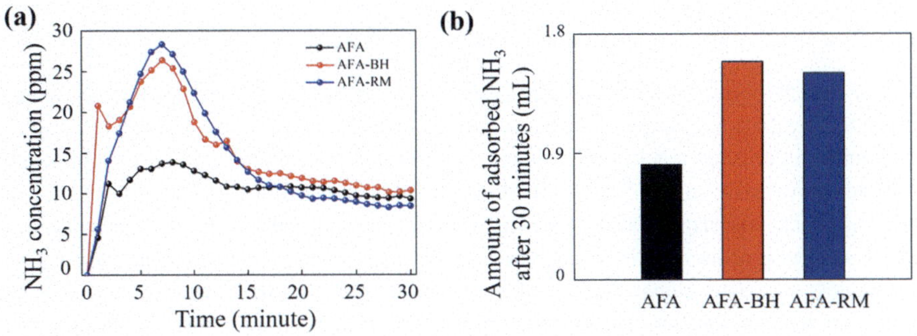

Fig. 5.3 **a** NH_3 adsorption removal rate of produced absorbents **b** the calculated capacity of NH_3 adsorption

penetration into the pores of the adsorbent material (refer to Table 5.1). This trend is consistent with a work completed by Huang et al., wherein they observed a decrease in the S_{BET} of the activated carbon (AC) from 1073 to 955 and 926 m^2/g after treatment with 1N and 12N nitric acid (HNO_3), respectively. Despite the reduced S_{BET} values, these modifications led to significantly higher capacities of NH_3, reaching 13.9 and 41.6 mg-NH_3/g-AC, attributed to chemisorption facilitated by acid modification. This marked an improvement from the 2.273 mg-NH_3/g-adsorbent capacity of unmodified AC [10]. Furthermore, the data shown in Table 5.1 demonstrate that the porosity and micropore structures (diameters less than 2 nm) of the material were promoted by steam activation. The micropore volume (V_{Micro}) of FA increased from 0.0035 to 0.0280 cm^3/g (eightfold increase) after steam activation. Additionally, S_{BET} of FA raised from 51.2 to 67.7 m^2/g (32% increase), while the average pore radius reduced from 3.10 to 1.45 nm. These results were consistent with the work of Zhou et al. (2018), who similarly summarized that microporous structures of material could be enhanced by steam activation [11].

Table 5.1 Porosity characteristics of material

Material	S_{BET} (m^2/g)	V_{total} (cm^3/g)	V_{Micro} (cm^3/g)	Average pore radius (nm)
FA	51.2	0.0629	0.0035	3.10
AFA	67.7	0.0817	0.0280	1.45
AFA-BH	47.2	0.0629	0.0208	1.98
AFA-RM	36.4	0.0630	0.0130	2.94

5.4 Conclusions

This research demonstrated the effectiveness of utilizing FA derived from waste generated by large-scale power plants, along with improvements using BH or RM. The study successfully achieved low-cost production without the need for chemical modifications, utilizing all raw byproducts. The findings emphasize the potential of utilizing waste materials for NH_3 removal, offering a cost-effective and environmentally friendly approach without the need for chemical modifications. This study also contributes to waste management solutions and highlights the importance of repurposing waste materials for environmental remediation purposes. The key findings of the study are summarized as follows.

1. Steam activation significantly enhanced the V_{Micro} and S_{BET} of FA, increasing from 0.0035 to 0.0280 cm^3/g (eightfold increase) and from 51.2 to 67.7 m^2/g (32% increase), respectively.
2. The combination of bagasse FA with 15 wt% bamboo hydrochar resulted in a capacity of 0.84 mL-NH_3/g-adsorbent. This high capacity was primarily due to the chemical adsorption of AFA/BH, predominantly facilitated by the functional groups present in BH, such as carboxyl and hydroxyl groups. While iron oxide contained in RM could enhance an NH_3 adsorption with a capacity of 0.75 mL-NH_3/g-adsorbent.

Acknowledgements The funding for this project was provided by the Office of the National Research Council of Thailand under the Fundamental Fund (Grant no. FF(KU)2.65), as well as by the Office of the Ministry of Higher Education, Science, Research, and Innovation, and Thailand Science Research and Innovation through the Kasetsart University Reinventing University Program 2022. T. Ketwong received a scholarship from the Department of Mechanical Engineering, Faculty of Engineering, Kasetsart University, Thailand.

References

1. X. Huang, J. Zhang, W. Zhang, G. Tang, Y. Wang, Atmospheric ammonia and its effect on PM2.5 pollution in urban Chengdu, Sichuan Basin, China. Environ. Pollut. **291**, 118195 (2021)
2. T. Ketwong, N. Cholwatthanatanakorn, L. Ding, H. Wibowo, C. Areeprasert, Utilization of bagasse fly ash for the production of low-cost ammonia adsorbents in poultry farm. Waste Manag. **172**, 347–357 (2023)
3. T. Ketwong, E.R. Halabaso, T.K.A. Nguyen, C. Areeprasert, R.-A. Doong, Comparative study on pilot-scale production of CuO-loaded activated biochar and hydrochar from oil-palm empty fruit bunches for high-performance symmetric supercapacitor application. J. Electroanal. Chem. **905**, 115970 (2022)
4. H. Wibowo, T. Ketwong, N. Cholwatthanatanakorn, L. Ding, C. Areeprasert, Production of bagasse fly ash-derived CO2 adsorbent by physical activation and by nitrogen-functionalization using hydrothermal treatment. Waste Manag. **177**, 66–75 (2024)
5. Z. Ren, B. Jia, G. Zhang, X. Fu, Z. Wang, P. Wang, L. Lv, Study on adsorption of ammonia nitrogen by iron-loaded activated carbon from low temperature wastewater. Chemosphere **262**, 127895 (2021)

6. W. Bai, M. Qian, Q. Li, S. Atkinson, B. Tang, Y. Zhu, J. Wang, Rice husk-based adsorbents for removing ammonia: kinetics, thermodynamics and adsorption mechanism. J. Environ. Chem. Eng. **9**, 105793 (2021)
7. Situation report of sugar cane cultivation Year 2021/22. Office of the Cane and Sugar Board, Thailand (2022)
8. Y. Li, C. Liu, Z. Luan, X. Peng, C. Zhu, Z. Chen, Z. Zhang, J. Fan, Z. Jia, Phosphate removal from aqueous solutions using raw and activated red mud and fly ash. J. Hazard. Mater. **137**, 374–383 (2006)
9. Y. Peng, J. Li, Ammonia adsorption on graphene and graphene oxide. A first-principles study. Front. Environ. Sci. Eng. **7**, 403–411 (2013).
10. C.-C. Huang, H.-S. Li, C.-H. Chen, Effect of surface acidic oxides of activated carbon on adsorption of ammonia. J. Hazard. Mater. **159**, 523–527 (2008)
11. J. Zhou, A. Luo, Y. Zhao, Preparation and characterisation of activated carbon from waste tea by physical activation using steam. J. Air Waste Manag. Assoc. **68**, 1269–1277 (2018)

Chapter 6
Synthesis of Activated Carbons from Corncob Biomass Waste as Electrode Materials for Supercapacitors

Chanatip Sungprasit, Kasidit Janbooranapinij, Nonthapat Kosacarn, Attanon Choomthi, and Gasidit Panomsuwan

Abstract Nowadays, activated carbons (ACs) derived from biomass waste are promising materials for various applications, particularly supercapacitors, due to their high surface area and porosity, low cost, natural abundance, power delivery performance, and electrochemical stability. In this work, ACs were synthesized from corncob waste through carbonization at 600°C followed by chemical activation using zinc chloride ($ZnCl_2$) at 800°C in various ratios: 1:1, 1:2, 1:3, and 1:4. The characterization results revealed that ACs exhibited an amorphous carbon phase without any impurities and had almost no chemical functional groups. The morphology of the ACs showed the development of a porous structure compared to raw corncob and non-activated samples. The specific surface area of ACs slightly increased from 417 to 450 m^2g^{-1} with the increasing $ZnCl_2$ content, mainly due to the development of micropores, with a smaller proportion of meso-macropores. The electrochemical properties of ACs were characterized using a three-electrode system in a 6 M KOH electrolyte via cyclic voltammetry (CV) and galvanostatic charge–discharge (GCD). The AC synthesized using a ratio of 1:4 showed the highest specific capacitance of 56 Fg^{-1} attributed to its highest specific surface area. Additionally, the capacitance retention remained at 92% after 5,000 cycles. This study provides valuable information on converting corncob waste into ACs as electrode materials in supercapacitor applications.

Keywords Corncob · Biomass · Activated carbons · Supercapacitors

---

C. Sungprasit · K. Janbooranapinij · N. Kosacarn · A. Choomthi · G. Panomsuwan (✉)
Department of Materials Engineering, Faculty of Engineering, Kasetsart University, Bangkok, Thailand
e-mail: gasidit.p@ku.ac.th

C. Sungprasit · K. Janbooranapinij · G. Panomsuwan
Special Research Units for Biomass Conversion Technology for Energy and Environmental Materials, Kasetsart University, Bangkok, Thailand

C. Areeprasert et al. (eds.), *Proceedings of the International Conference on Sustainable Energy and Environmental Technology for Circular Economy*, Springer Proceedings in Energy, https://doi.org/10.1007/978-981-96-2306-8_6

6.1 Introduction

Currently, energy development encompasses various forms, such as solar and wind energy, which are driving the advancement of energy storage systems like rechargeable batteries and supercapacitors [1]. These devices allow for storing energy based on the application and specific requirements. Among the diverse energy storage devices, batteries and fuel cells are notable for their ability to store substantial amounts of energy, though they deliver it at a slow rate [2]. Supercapacitors offer distinct advantages, such as high-power density, long cycle life (> 100,000 cycles), and enhanced safety. They are utilized in diverse fields, including powering hybrid and electric vehicles, smart grids, UPS systems, and a wide range of consumer electronics [3]. The energy storage mechanisms of supercapacitors are typically categorized into three types: (i) electrochemical double-layer capacitors (EDLCs), (ii) pseudocapacitors, and (iii) hybrid supercapacitors [4]. The choice of electrode type plays a crucial role in determining capacitance and charge storage efficiency, influenced by several factors such as specific surface area, pore structure, surface chemistry, and electrical conductivity [5]. Pseudocapacitive materials often exhibit higher specific capacitance than EDLCs due to their higher energy density. However, EDLCs with carbon-based electrodes offer superior cycle stability, and rate capability, and can operate at higher charge/discharge rates with a longer lifetime [6].

Activated carbons (ACs) are widely favored as electrode materials for supercapacitors due to their unique properties, including low cost, large specific surface area, microporosity, high conductivity, and chemical stability [7]. They can be synthesized through the carbonization and activation processes. These processes yield ACs rich in micropores, which are highly suitable for supercapacitor applications [8]. In recent years, plants, abundant in cellulose, hemicellulose, protein, lignin, starch, and other carbonaceous substances, have served as a viable source for synthesizing ACs [9]. Numerous studies have demonstrated that biomass-derived ACs offer diverse sources, low production costs, and environmental friendliness. Their porous structures provide active sites crucial for effective electrode materials in supercapacitors. The structural variability inherent in biomass results in ACs with a range of morphologies, structures, porosities, and electrochemical properties [10].

In this study, ACs were synthesized by first carbonizing at 600°C followed by activation using $ZnCl_2$ as the activating agent at 800°C. The resulting ACs were characterized comprehensively by various instruments to investigate their morphology, porosity, structure, and chemical functionality. Furthermore, the electrochemical charge storage capabilities of ACs were evaluated using a three-electrodes system at room temperature in a 6 M KOH electrolyte.

6.2 Experimental

6.2.1 Materials

Zinc chloride ($ZnCl_2$, 98% purity) was purchased from DAEJUNG Co., Ltd. Ethanol (C_2H_2OH, 99.9% purity) and isopropanol (C_3H_8O, 99.8% purity) were purchased from RCI Labscan Ltd. Hydrochloric acid (HCl, 37%) was purchased from Qrëc Chemical, Co., Ltd. Nafion® solution (5% wt in a mixture of water and alcohols) was purchased from Fuel Cell Earth Co., Ltd. All chemicals used in this study were of analytical grade.

6.2.2 Synthesis of Activated Carbons (ACs)

Corncob waste was initially dried at 105°C for 12 h and then ground using a grinding machine. The resulting dried and ground corncob powder was carbonized at 600°C for 3 h under 0.2 L min^{-1} Ar flow, yielding non-activated carbon (non-AC). For the activation process, $ZnCl_2$ was dissolved in water and mixed with the non-AC under magnetic stirring for 1 h, followed by drying at 105°C for 24 h. The weight ratio of the non-AC to $ZnCl_2$ varied at 1:1, 1:2, 1:3, and 1:4. Subsequently, the mixed samples were subjected to a tube furnace for activation at 800°C for 1 h under a 0.2 L min^{-1} Ar flow. After activation, the products were washed with 1 M HCl and deionized water until the filtrate reached a pH of 7, followed by drying at 105°C for 24 h. The AC samples were designated based on the ratio of non-AC:$ZnCl_2$ as AC1:1, AC1:2, AC1:3, and AC1:4 (Fig. 6.1).

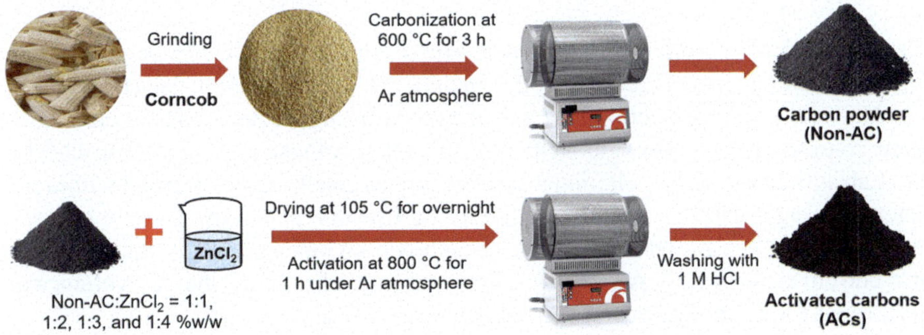

Fig. 6.1 Schematic illustration showing the preparation of ACs from corncob waste

6.2.3 Characterization

Various analytical techniques were employed to investigate the properties of ACs. The morphology was examined using a Hitachi SU3500 scanning electron microscope. N_2 adsorption–desorption isotherms of the activated carbons were recorded at—196°C using a Micromeritics 3Flex surface characterization analyzer to evaluate their specific surface area and pore structure. Prior to the measurement, the samples were degassed at 150°C for 12 h under vacuum conditions. The phase structure of the ACs was verified using a PANalytical Empyrean X-ray diffractometer with Cu K_α radiation. The chemical functional group was analyzed with a Bruker Alpha-E Spectrometer in the transmittance mode, within the wavenumber range of 4000–500 cm^{-1}.

6.2.4 Preparation of Working Electrode

The ACs (5 mg) were first dispersed in a mixture comprising 475 μL ultrapure water, 475 μL isopropanol, and 50 μL Nafion. The mixture was then sonicated for 30 min to achieve a homogeneous suspension. The glassy carbon (GC) electrode was thoroughly polished on a polishing pad with diamond and alumina polishing pads. After polishing, the electrode was rinsed with deionized water and air-dried. Subsequently, 3 μL of the prepared suspension was dropped onto the GC electrode and air-dried.

6.2.5 Electrochemical Measurements

Electrochemical measurements were performed using a three-electrodes system in a 6 M KOH electrolyte. Three electrodes included a reference electrode (Hg/HgO in 1 M NaOH), a counter electrode (Pt wire), and a working electrode. These electrodes were connected by a Biologic VSP potentiostat/galvanostat station controlled by EC-Lab software. Cyclic voltammetry (CV) measurement was performed within the potential range of 0 to –1 V at a scan rate of 100 mV s^{-1}. Galvanostatic charge–discharge (GCD) was conducted in the same potential range as the CV measurement at a current density of 1 A g^{-1}. To assess stability, charge–discharge cycling was performed for 5,000 cycles at a current density of 10 A g^{-1}.

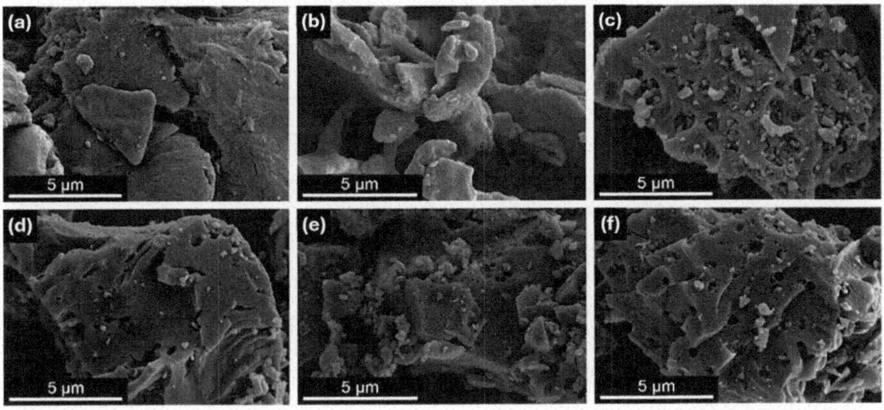

Fig. 6.2 SEM images: **a** corncob, **b** non-AC, **c** AC1:1, **d** AC1:2, **e** AC1:3, and **f** AC1:4

6.3 Results & Discussion

6.3.1 Morphology

The morphology of corncob, non-AC, and ACs was investigated by scanning electron microscopy (SEM), as displayed in Fig. 6.2. The surface structure of both corncob and non-AC showed a ruptured surface without visible pores. However, after activation with $ZnCl_2$, a significant increase in surface porosity was observed in the AC samples compared to the corncob and non-AC samples.

6.3.2 Surface Area and Porosity Analysis

The N_2 adsorption–desorption isotherms of corncob and non-AC exhibited a type II feature (Fig. 3a), indicating a non-porous structure with the presence of macropores. In contrast, the AC samples displayed a combination of type I and IV features. The initial abrupt adsorption in the isotherms corresponds to type I, indicative of micropore structures, while the hysteresis loop during desorption corresponds to type IV, suggesting mesopores with narrow slit-like shapes, as shown in Fig. 3b. The specific surface areas determined using the Brunauer–Emmett–Teller (BET) method were 417, 432, 447 and 450 $m^2\ g^{-1}$ for AC1:1, AC1:2, AC1:3, and AC1:4, respectively, which was much higher than corncob (1.32 $m^2\ g^{-1}$) and non-AC (1.16 $m^2\ g^{-1}$). The slight increase in specific surface area was attributed to the formation of additional micropores, while the proportion of meso- and macropores remained relatively unchanged. The total pore volume and average pore diameter of ACs are listed in Table 6.1. These findings indicate that the increasing $ZnCl_2$ content had a

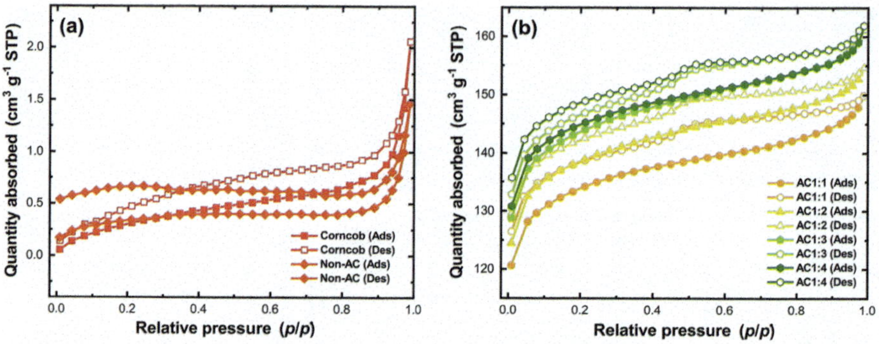

Fig. 6.3 N_2 adsorption–desorption isotherms of **a** corncob and non-AC and **b** AC samples (AC1:1, AC1:2, AC1:3, and AC1:4)

Table 6.1 The parameters obtained from N_2 adsorption–desorption analysis

Samples	S_{BET}(m^2 g^{-1})	S_{micro}(m^2 g^{-1})	$S_{meso\text{-}macro}$(m^2 g^{-1})	Pore volume (cm^3g^{-1})	Pore diameter (nm)
AC1:1	417	349	68	0.228	3.51
AC1:2	432	361	71	0.237	3.54
AC1:3	447	373	74	0.244	3.69
AC1:4	450	382	68	0.244	3.82

slight influence on creating the porous structure and enhancing the surface area of ACs.

6.3.3 *Phase Structure*

The phase structure of corncob, non-AC, and ACs was evaluated by X-ray diffraction (XRD), as shown in Fig. 4a. The XRD pattern of corncob exhibited peaks at 16.0°, 22.5°, and 34.5°, corresponding to the cellulose crystalline structure [11]. For non-AC, the peaks associated with crystalline cellulose were absent, replaced by broad peaks of amorphous carbon at 24.3° and 43.9°, corresponding to the (002) and (101) planes, respectively. Additionally, small peaks from impurity elements of corncob were also observed. After activation with $ZnCl_2$, the XRD patterns of the ACs exhibited only amorphous carbon peaks, indicating high carbon purity and complete removal of inorganic salts from the activated products.

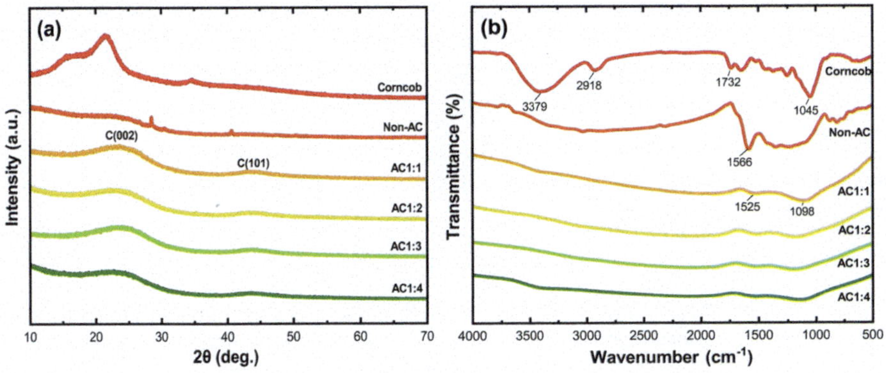

Fig. 6.4 **a** XRD patterns and **b** FTIR spectra of corncob, non-AC, and ACs (AC1:1, AC1:2, AC1:3, and AC1:4)

6.3.4 Chemical Functional Groups

The chemical functional groups were analyzed by Fourier-transform infrared spectroscopy (FTIR), as shown in Fig. 4b. The FTIR spectrum of the corncob showed absorption peaks at 3379, 2918, 1732, and 1045 cm^{-1}, corresponding to the O–H, C–H, C = O, and C–O bonds in the lignocellulosic structure, respectively [12]. After carbonization, most of these absorption peaks diminished due to the decomposition of the lignocellulosic structure and the removal of O and H at high temperatures. A peak emerging at 1566 cm^{-1} was assigned to the C = C stretching. After activation, the FTIR spectra displayed a nearly featureless profile, with only weak intensity peaks remaining. These included the O − H stretching at 3380 cm^{-1}, C = C stretching at 1525 cm^{-1}, and C − O stretching at 1098 cm^{-1}, which are the typical characteristics of ACs [13, 14].

6.3.5 Electrochemical Charge Storage Performance

The electrochemical properties of ACs for supercapacitor applications were evaluated in a 6 M KOH solution using CV and GCD measurements. Figure 5a illustrates the CV curves of the ACs recorded from 0 to – 1 V at a scan rate of 100 mV s^{-1}. The area of CV curves varied proportionally with the surface area of the ACs, with AC1:4 exhibiting the largest area, indicating the highest capacitance among all tested ACs. Comparative GCD curves of ACs at a current density of 1 A g^{-1} are shown in Fig. 5b, demonstrating non-linear behavior. The specific capacitance (C_s) can be calculated using the integrated area under the discharge curve following equation [15]:

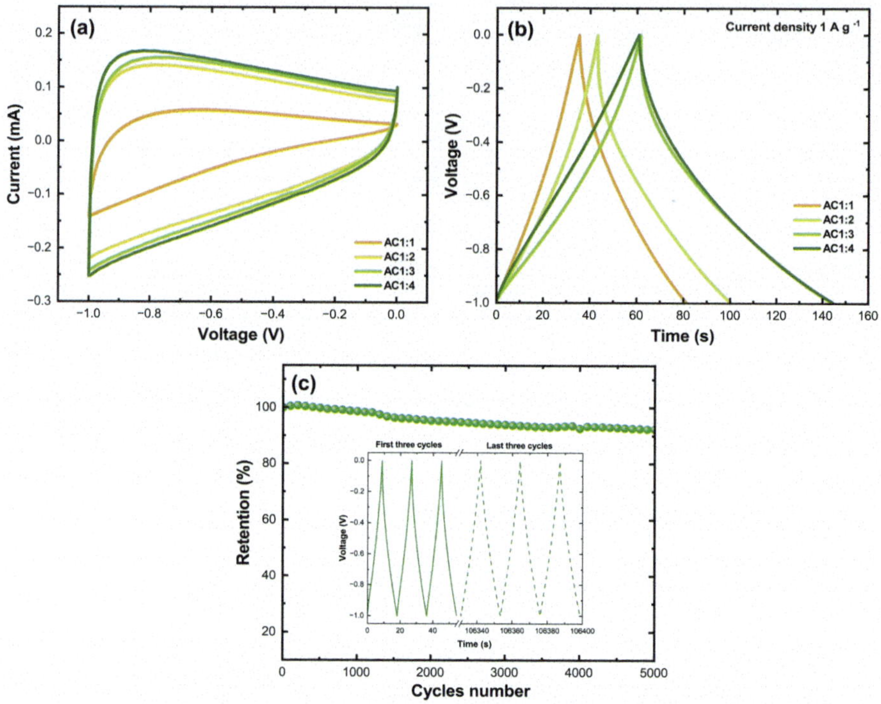

Fig. 6.5 **a** Comparative CV curves at a scan rate of 100 mV s^{-1} and **b** GCD curves at a current density of 1 A g^{-1}, and **c** capacitance retention of AC1:4 electrode at a current density of 10 A g^{-1} over 5,000 cycles

$$C_s = \frac{2I}{m(V_f - V_i)^2} \int_{V_i}^{V_f} V dt, \tag{6.1}$$

where I is the applied current (A), m is the mass of AC, and V_f and V_i are the final and initial values of the potential range (V), respectively. At a current density of 1 A g^{-1}, C_s values were determined as 32, 37, 53, and 56 F g^{-1} for AC1:1, AC1:2, AC1:3, and AC1:4, respectively. The highest C_s value of AC1:4 can be attributed to its highest specific surface area, facilitated by increased micropore creation due to high $ZnCl_2$ content. Moreover, AC1:4 was selected for a stability test via 5,000 charge–discharge cycles at a current density of 10 A g^{-1} (Fig. 5c). The capacitance retention remained at 92% after 5,000 cycles, assessing its excellent stability and suitability for long-term supercapacitor applications.

6.4 Conclusion

ACs were successfully synthesized via a two-step process, including carbonization at 600ºC followed by chemical activation at 800ºC using varying ratios of $ZnCl_2$ (1:1, 1:2, 1:3, and 1:4). Characterization results showed that the ACs exhibited an amorphous structure without impurity phases, and the chemical functional groups were effectively removed after activation at high temperatures. The increase in $ZnCl_2$ content slightly influenced the creation of additional micropores and enhanced the surface area of ACs by 8%. Electrochemical measurements revealed that AC1:4 exhibited the highest specific capacitance of 56 F g^{-1} among all tested ACs. This was attributed to its highest specific surface area contributed by an increased presence of micropores. After a stability test for 5,000 charge–discharge cycles, the retention remained at 92%, demonstrating its excellent stability for prolonged operational use. Although the specific surface area and capacitance values were relatively low compared to reported ACs in the literature, this study highlights the potential of utilizing corncob waste as a precursor for AC production in supercapacitor electrodes. Future research will focus on optimizing other factors and refining the activation conditions to enhance the specific surface area and porosity of these ACs, making them more suitable for practical applications in supercapacitors.

Acknowledgements This work was financially supported by the Kasetsart University Research and Development Institute (KURDI, grant no. FF(KU) 51.67) and the Innovation Development Scholarship for Master's students from the Faculty of Engineering, Kasetsart University (67/05/MAT/Innovation). The authors would also like to thank the Department of Materials Engineering and the Materials Innovative Center at Kasetsart University for their support in providing facilities and instruments.

References

1. A. Mardani, A. Jusoh, E.K. Zavadskas, F. Cavallaro, Z. Khalifah, Sustainable and renewable energy: an overview of the application of multiple criteria decision making techniques and approaches. Sustainability **7**(10), 13947–13984 (2015)
2. J. Mitali, S. Dhinakaran, A.A. Mohamad, Energy storage systems: a review. Energy Storage Sav. **1**, 166–216 (2022)
3. A. Afif, S.M.H. Rahman, A.T. Azad, J. Zaini, M.A. Islan, A.K. Azad, Advanced materials and technologies for hybrid supercapacitors for energy storage–a review. J. Energy Storage **25**, 100852 (2019)
4. S. Sharma, R. Chand, Supercapacitor and electrochemical techniques: a brief review. Results Chem. **5**, 100885 (2023)
5. B. Arumugam, G. Mayakrishnan, S.K.S. Manickavasagam, S.C. Kim, R. Vanaraj, An overview of active electrode materials for the efficient high-performance supercapacitor application. Crystals **13**(7), 1118 (2023)
6. H.W. Park, K.C. Roh, Recent advances in and perspectives on pseudocapacitive materials for supercapacitors-a review. J. Power. Sources **557**, 232558 (2023)

7. L. Luo, Y. Lan, Q. Zhang, J. Deng, L. Luo, Q. Zeng, H. Gao, W. Zhao, A review on biomass-derived activated carbon as electrode materials for energy storage supercapacitors. J. Energy Storage **55**, 105839 (2022)
8. M. Chen, X. Kang, T. Wumaier, J. Dou, B. Gao, Y. Han, G. Xu, Z. Liu, L. Zhang, Preparation of activated carbon from cotton stalk and its application in supercapacitor. J. Solid State Electrochem. **17**(4), 1005–1012 (2013)
9. J. Deng, T. Xiong, H. Wang, A. Zheng, Y. Wang, Effects of cellulose, hemicellulose, and lignin on the structure and morphology of porous carbons. ACS Sustain. Chem. & Eng. **4**(7), 3750–3756 (2016)
10. B. Lesbayev, M. Auyelkhankyzy, G. Ustayeva, M. Yeleuov, N. Rakhymzhan, A. Maltay, Y. Maral, Recent advances: biomass-derived porous carbon materials. S. Afr. J. Chem. Eng. **43**, 327–336 (2023)
11. J. Gong, J. Li, J. Xu, Z. Xiang, L. Mo, Research on cellulose nanocrystals produced from cellulose sources with various polymorphs. RSC Adv. **7**, 33486 (2017)
12. F. Abdallah, E.K. Arthur, K. Mensah-Darkwa, E. Gikunoo, S.A. Baffour, B.A. Agamah, M.A. Nartey, F.O. Agyemang, Electrochemical performance of corncob-derived activated carbon-graphene oxide and TiO_2 ternary composite electrode for supercapacitor applications. J. Energy Storage **68**, 107776 (2023)
13. S. Wanprakhon, P. Pattaraphutanon, P. Borvornsudhasin, N. Choocherd, Chemical and surface properties of activated carbon from banana peel by dry chemical activation. J. Mater. Sci. Appl. Energy **10**(3), 96–105 (2021)
14. I. Ozdemir, M. Sahin, R. Orhan, M. Erdem, Preparation and characterization of activated carbon from grape stalk by zinc chloride activation. Fuel Process. Technol. **125**, 200–206 (2014)

Chapter 7
A Study on the Use of Food-Grade and Non-food Grade Reagents in the Extraction of Cellulosic Fibers from Bamboo Species in Baguio City, Philippines

Jennifer Bandao-Antonio

Abstract Cellulosic fibers were extracted from bolo (***Gigantochloa bolo***) culm and running bamboo (***Phyllostachys aurea***) branches grown in Baguio City, Philippines using separate pre-treatment and processing reagents. Food grade reagents such as acetic acid and hydrogen peroxide were used to further process the fibers after $NaHCO_3$ pre-treatment and non-food grade reagents such as hydrochloric acid were used to further refine another set of NaOH pre-treated fiber samples. Fourier Transform Infrared (FTIR) analysis revealed that the use of food grade reagents in the fiber extraction process was effective in removing surface impurities as shown in the general decrease in percentage absorbance (%A) of functional groups associated to hemicellulose and lignin. This is also observed in the increasing crystallinity index (CI) calculated from x-ray diffractogram for six days NaOH pre-treatment, six days $NaHCO_3$ pre-treatment, and 10 days $NaHCO_3$ pre-treatment of bolo culms. However, in terms of fiber diameter NaOH pre-treatment was best in decreasing fiber diameter in running bamboo stem as shown in the optical microscopy results. The scanning electron microscopy (SEM) displayed a smoother and more organized structure for NaOH pre-treated bamboo fibers. In general, volumetric analysis using potassium dichromate ($K_2Cr_2O_4$) and ferrous ammonium sulfate ($FeH_8N_2O_8S_2$) in a reduction–oxidation reaction confirmed the presence of alpha(α) cellulose in the bamboo samples as evidenced by a purple end-point. Further investigation of processing parameters using food grade or eco-friendly reagents can be done as this initial investigation shows the potential of food-grade reagents in the extraction of cellulosic fibers from bamboo.

J. Bandao-Antonio (✉)
Department of Physical Sciences, College of Science, Military Cut-Off, University of the Philippines Baguio, Governor Pack Road, 2600 Baguio City, Philippines
e-mail: jbantonio4@up.edu.ph

C. Areeprasert et al. (eds.), *Proceedings of the International Conference on Sustainable Energy and Environmental Technology for Circular Economy*, Springer Proceedings in Energy, https://doi.org/10.1007/978-981-96-2306-8_7

Keywords Cellulosic bamboo fibers · Alkaline pre-treatment · Food grade reagents

7.1 Introduction

The usual process of fiber extraction from plants includes mechanical crushing, steam explosion techniques, use of biological enzymes and water retting techniques. Chemical pre-treatment methods such as use of NaOH, HCl, H_2SO_4 are also considered in combination with mechanical crushing or steam explosion. While chemical pre-treatment methods are proven to be effective, the reagents are toxic to human health and environment with large-scale use. The use of food grade reagents such as sodium bicarbonate ($NaHCO_3$) and acetic acid (CH_3COOH) in vinegar are cheaper and safer to work on. These are also widely available and easy to purchase.

The use of ecofriendly reagents such as ethanol, acetic acid, and hydrogen peroxide was done by Rehman et al. in 2018 and they were able to successfully isolate cellulose from eucalyptus plant with 73% crystallinity [1]. 12 wt% of $NaHCO_3$ treatment for 120 h soaking time on coir fibers were also performed. The resulting fibers increased in tensile strength, observance of lesser micropores thus decreased local stress, and increased crystallinity index [2]. $NaHCO_3$ treatment was also reported to be effective and efficient in removing the surface impurities and hemicellulose content of jute/hemp/flax bidirectional mats. 9% of NaHCO3 solution with 96-h soaking time was concluded to be effective [3]. Untreated Napier grass fiber strand was also compared to acetic acid treated samples. Acetic acid at 15 wt% and 2 h soaking time was found to have increased crystallinity index and thermal stability of the fiber samples [4]. In another study, 20% $NaHCO_3$ solution at 120 h soaking time was effective on flax fiber extraction. The extracted fibers' tensile strength was reported to increase tensile strength by 43% and increased young's modulus at 81% [5]. Fiber treated with $NaHCO_3$ when used to reinforce polymer composites resulted to better properties. The tensile and flexural properties of polymer composites reinforced with flax, jute, and sisal fibers were also observed to show significant increase as compared to composites reinforced with NaOH treated fibers [6]. The study made by dos Santos et al. has also shown that laminates made with polyester polymer and sisal fiber treated with Na_2CO_3 for 96 h have remarkable increase in tensile and flexural modulus as well as reduced impact resistance [7].

The extraction of fibers from bamboo were also experimented on due to bamboo fibers' potential use in manufacturing industries. Bamboo fibers can replace petrochemical synthetic fibers due to its superior mechanical properties, sustainability and versatility. Zheng et al. used the acetic acid ball milling technology as pre-treatment to moso bamboo powder to substitute traditional fiber board and particle board as furniture materials. The acetic acid ball-milled moso bamboo powder was more compact resulting to a superior mechanical property in terms of tensile and specific strength [8]. Steam explosion, mechanical techniques, and chemical extraction using alkali or NaOH were used to extract bamboo fibers. It was found that the chemical treatment increased the interfacial shear strength (IFSS) of bamboo with polyester and polypropylene composites. It was suggested by the researchers that airing of waste water from the alkaline extraction process using CO_2 gas can reduce the pH of waste water from 10.8 to 7.0 within 110 min aeration time [9]. In another study, it was also discovered that alkali or NaOH treatment could remove lignin from bamboo fibers thus improving the interfacial adhesion of bamboo fibre-reinforced composites [10].

This study on the use of eco-friendly reagents to extract fibers from bolo (***Gigantochloa bolo***) culm and running bamboo (***Phyllostachys aurea***) branches, if proven effective would be an ecologically friendly means of fiber extraction. The inorganic reagents sodium hydroxide (NaOH), sulfuric acid (H_2SO_4), and hydrochloric acid (HCl) used in the processing of plant fiber sources for extraction purposes are both corrosive and toxic to humans who are often exposed to these. The disposal of waste water from the extraction process poses problem to the environment. The use of food grade reagents such as sodium bicarbonate ($NaHCO_3$) /baking soda and acetic acid (CH_3COOH) in vinegar would be safer to handle and dispose aside from its availability. If proven to be effective, then bamboo fibers extracted using food grade reagents can be incorporated in food bio- packaging applications, edible bioplastics, textile and composite fabrication in industries—aeronautics, automotive, furniture. However, the mechanical properties of extracted fibers, applications in composite and textile fabrication still need to be explored and experimented on as a result of this groundwork study.

7.2 Materials and Methodology

Two species of bamboo, the bolo (***Gigantochloa bolo***) culm and running bamboo (***Phyllostachys aurea***) branches were used as a source of extract fibers. These are locally grown in Baguio City, Philippines.

7.2.1 Bamboo Fiber Extraction Process

Five to seven grams of chopped and sun-dried bamboo samples from bolo culms and running bamboo branches were pre- treated with 1 M NaOH for 6 days, mechanically crushed in a mortar and pestle then acid hydrolyzed with 1 M HCl at 70–80 °C for 2 h. The acid hydrolyzed samples were again soaked in 1 M NaOH for 2 h at 70–80 °C followed by hydrogen peroxide treatment, 30% v/v solution, for 12 h. The chemically treated fibers were subjected to a high-pressure defibrillation for 1 min using an Oster osterizer to separate the lignin, hemicelluloses, and the crystalline portion of the fiber. The fiber extraction of five to seven grams chopped and sun-dried bamboo samples from bolo culms and running bamboo branches using food grade reagents employed the same method stated using food grade reagents such as sodium bicarbonate ($NaHCO_3$), acetic acid (CH_3COOH), and hydrogen peroxide (H_2O_2).

7.2.2 *Characterization Techniques for the Extracted Bamboo Fiber*

FTIR Analysis

Fourier Transform Infrared with Attenuated Total Reflectance (FTIR-ATR) analysis using Agilent Cary 630 FTIR was done to determine the percentage absorbance (%A) of functional groups at 890–900 cm^{-1} due to β-glycosidic linkage by both cellulose and hemicellulose [11]. The %A at 1,637 cm^{-1} (C=C stretching) and 1428 cm^{-1} (aromatic ring stretching) both attributed to lignin was considered in the spectra. The contribution of hydroxyl (OH) groups from cellulose, hemicellulose, and lignin were also determined at 3,200 to 3,400 cm^{-1}.

SEM Analysis

The Scanning Electron Microscope (SEM) used was JEOL JSMIT500HR/LA. It was employed in the study to further understand the surface morphology of the extracted fiber strands from bamboo fiber. The samples were coated with platinum and processed using a landing voltage of 15.0 kV at 1,000 X magnification under high vacuum.

XRD Analysis

X-ray diffraction (XRD) analysis was performed to know the % Crystallinity Index (%CI) of the extracted bamboo fibers. The Shimadzu Lab X XRD-6100 X-ray Diffractometer was the brand of the machine, programmed at θ Range: 5°–60° with the rate of 1.0° per minute. Fiber samples were pressed in the sample holder. The CI was calculated from the XRD data using the Segal Method [12] as seen in Eq. 7.1:

$$CI = \frac{I(002) - I(am)}{I(002)} \times 100 \tag{7.1}$$

where:

CI = % Crystallinity Index.
I(002) = peak intensity of the crystalline portion.
I(am) = peak intensity of the amorphous portion.

Optical Microscopy for Fiber Fineness

Optika compound microscope, B-190 series at 50X magnification was used to determine fiber fineness. The external fiber diameter was measured using a microscope micrometer placed on the stage. Twenty random fiber strands were picked and placed on the stage for measurement.

Volumetric Analysis of Alpha (A) Cellulose Content

Bamboo fibers was reported to contain 40–50% α cellulose [13]. In the study, the extracted bamboo fibers were qualitatively tested of its α cellulose content using a reduction–oxidation reaction of potassium dichromate ($K_2Cr_2O_4$) and ferrous ammonium sulfate ($FeH_8N_2O_8S_2$). 25.0 mL of the previously prepared cellulose filtrate and 10.0 mL of 0.5N potassium dichromate solution was placed into a 250 mL flask, 50 mL of concentrated H_2SO_4 was cautiously added. The solution remained hot for 15 min then cooled to room temperature with the addition of 50 mL distilled water. 2–4 drops of Ferroin indicator were added then titrated with 0.1N ferrous ammonium sulfate solution to a purple color. The procedure was adopted from Technical Association of the Pulp and Paper Industry, TAPPI, 1999 [14].

7.3 Results and Discussion

The extraction of bamboo fibers from culm of bolo (***Gigantochloa bolo***) and branches of running bamboo (***Phyllostachys aurea***) were successfully performed using 6 days NaOH pre-treatment followed by additional procedures described in the methods portion of this paper which is identified as use of non-food grade reagents. Apparently, the 6 days and 10 days $NaHCO_3$ pre-treatment followed by extraction procedures identified as use of food grade reagents were also observed to be a success except for some processing parameters that need further improvement.

Thorough inspection of the FTIR spectra reveals an increase in the percentage absorbance (% A) of hydroxyl groups, lignin, and glycosidic linkage by both cellulose and hemicellulose for 6 days NaOH pre-treatment and 6 days $NaHCO_3$ pre-treatment. The -OH groups contributed by cellulose, hemicellulose, and lignin did not decrease with $NaHCO_3$ pre-treatment which means that surface impurities were not removed by $NaHCO_3$ pre-treatment for 6 days. Yet, at 10 days $NaHCO_3$ pre-treatment time there was a decrease in the % A of -OH groups and β-glycosidic linkage by cellulose and hemicellulose. This clearly explains the need for longer $NaHCO_3$ pre-treatment time of the bamboo culms (***Gigantochloa bolo***) to effectively remove surface impurities such as lignin and hemicellulose thereby extracting more cellulose from the fiber bundle. It is also suggested to chop the bamboo culm (***Gigantochloa bolo***) into thinner pieces so that $NaHCO_3$ will easily penetrate into the fiber bundles that will eventually cause removal of hemicelluloses and lignin.

In plants, the fiber bundle functions for structural support making the plant stronger. When fibers are used as a material in composites and textiles, hemicelluloses and lignin are weaker and decrease interfacial adhesion to other bio-polymers therefore the more crystalline cellulose is of interest to material scientists. The use of a basic solution such as NaOH and $NaHCO_3$ in fiber pre-treatment breaks the hydroxyl (-OH) bonds that holds cellulose, hemicelluloses, and lignin together. Depolymerization occurs, producing a more crystalline fiber [15].

A general decrease in the % A of functional groups related to cellulose, hemicelluloses, and lignin for running bamboo (***Phyllostachys aurea***) branches from 6 days NaOH pre-treatment to 6 days $NaHCO_3$ and 10 days $NaHCO_3$ pre-treatment is presented in Table 7.1. The decrease in %A means that the 6 days and 10 days pre-treatment time for running bamboo (***Phyllostachys aurea***) branches was effective in removing hemicelluloses and lignin from the fiber bundle, thus more cellulose is obtained. While NaOH pre-treatment is proven to be good in extracting cellulosic fibers, the best pre-treatment time with the use of $NaHCO_3$ on running bamboo (***Phyllostachys aurea***) is 10 days exemplified by the remarkable decrease in %A of functional groups related to cellulose, hemicellulose, and lignin.

SEM analysis reveals a more fibrous structure for NaOH pre-treated bolo (***Gigantochloa bolo***) culm and running bamboo (***Phyllostachys aurea***) fibers. Figure 7.1 shows the SEM micrographs of the extracted bamboo fibers. It can be seen that 6 days pre-treatment using 1 M NaOH for both bamboo culm (***Gigantochloa bolo***) and running bamboo (***Phyllostachys aurea***) branches was effective for producing a more fibrous and less damaged bamboo fiber. Bolo culm fiber pre-treated with 8 wt% $NaHCO_3$ for 6 days and 10 days have torn and rougher structure. This might mean that the pre-treatment time and concentration might not be enough to produce a more fibrous structure. However, for running bamboo (***Phyllostachys aurea***) branches the 6 days pre-treatment time with $NaHCO_3$ starts to produce more longitudinal structure and at 10 days pre-treatment fibers are being separated from the fiber bundle. This means that $NaHCO_3$ pre-treatment is effective in separating the bamboo fiber bundles into fiber strands. Perhaps a longer pre-treatment time is still needed to produce a finer fiber structure.

Table 7.1 Percentage Absorbance (%A) identified from FTIR spectra for running bamboo (***Phyllostachys aurea***) branches

Running bamboo fiber from branches	%A at specified pre-treatment time			
Wavenumber (cm^{-1})	Characteristics	6 days NaOH	6 days $NaHCO_3$	10 days $NaHCO_3$
3,200–3,400	Hydroxyl group contributed by cellulose, hemicellulose and lignin	14.11	12.558	8.636
1,630–1,640	C = C stretching of lignin	10.213	9.556	5.703
1,420–1,430	Lignin aromatic stretching	12.859	10.528	7.023
890–900	β-glycosidic linkage by both cellulose and hemicellulose	19.191	15.874	10.386

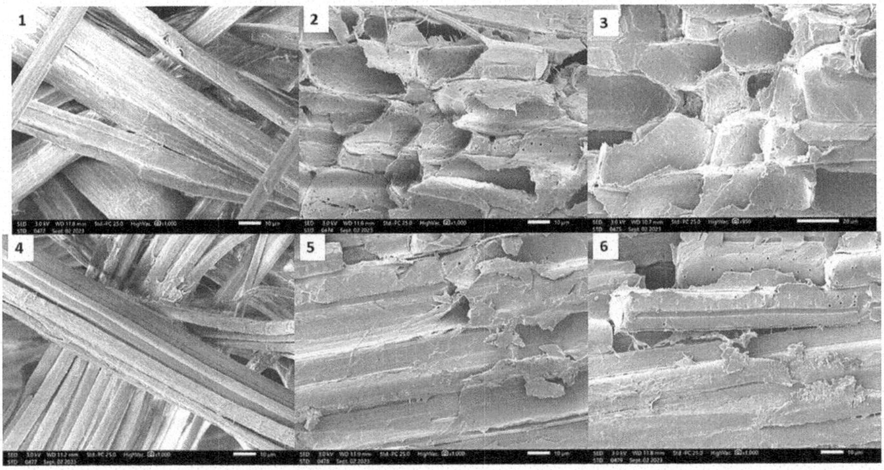

Fig. 7.1 These are SEM micrographs of the extracted fibers for 1—bolo culm pre-treated with 1M NaOH for 6 days, 2—bolo culm pre-treated with 8 wt% $NaHCO_3$ for 6 days, 3—bolo culm pre-treated with $NaHCO_3$ for 10 days, 4—running bamboo fibers pre-treated with 1M NaOH for 6 days, 5—running bamboo fibers pre-treated with 8 wt% $NaHCO_3$ for 6 days, and 6—running bamboo fibers pre-treated with 8 wt% $NaHCO_3$ for 10 days. The micrographs were taken at 1000 X magnification at a reference length scale of 10 um

X-ray diffraction analysis gives an idea of the crystallinity index (CI) of the extracted fibers. Figure 7.2 communicates the calculated values from the diffractograms of the bamboo samples. It can be seen that the most crystalline fiber was extracted from running bamboo (***Phyllostachys aurea***) branches pre-treated with NaOH at 6 days soaking time. For bolo (***Gigantochloa bolo***) culm, the most cellulosic fiber came from the 10 days $NaHCO_3$ pre-treated sample.

Fiber fineness is measured using the external diameter of the extracted fiber using an optical microscope. The fiber diameter on the average, 78.75 ± 70 μm for extracted bolo (***Gigantochloa bolo***) culm fibers were finer for NaOH pre-treated sample. This is similar to the extracted running bamboo (***Phyllostachys aurea***) fiber using NaOH pre-treatment with an average fiber diameter of 17.88 ± 14 μm. However, it can be seen that longer pre-treatment time for $NaHCO_3$ soaked running bamboo (***Phyllostachys aurea***) is needed to improve fiber fineness since the diameter of the 6 days pre-treated sample and 10 days pre-treated sample decreased from 70.38 ± 84 μm to 52.00 ± 87 μm. This is similar to the work done on alkali pre-treated kenaf fibers where increasing soaking time causes reduction in fiber diameter due to the degradation of impurities such as lignin with alkalinization [16]. Comparing the external fiber diameter of extracted bamboo fibers in this study to the study done by other researchers with a value of 107.9 μm using similar pre-treatment [17], it can be said that most of the extracted fibers in this study are finer at 78.75, 17.88, 70.38, and 52.00 μm (Table 7.2).

Alpha (α) cellulose is the undegraded form of cellulose. Volumetric method of analysis using potassium dichromate ($K_2Cr_2O_7$) and ferrous ammonium sulfate

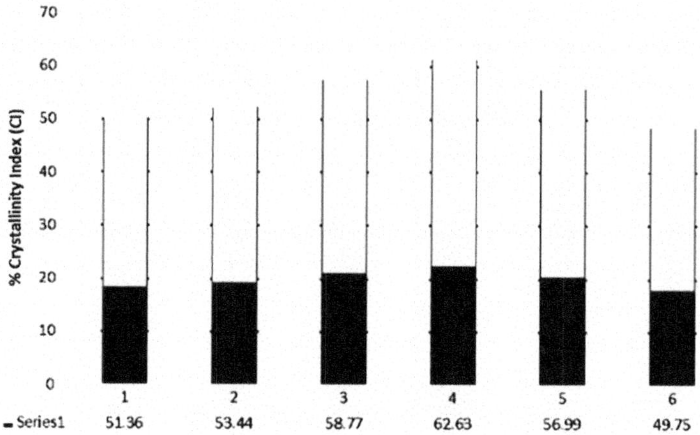

Fig. 7.2 These are the calculated CI for 1—bolo culm pre-treated with 1M NaOH for 6 days, 2—bolo culm pre-treated with 8 wt% $NaHCO_3$ for 6 days, 3—bolo culm pre-treated with $NaHCO_3$ for 10 days, 4—running bamboo fibers pre-treated with 1M NaOH for 6 days, 5—running bamboo fibers pre-treated with 8 wt% $NaHCO_3$ for 6 days, and 6—running bamboo fibers pre-treated with 8 wt% $NaHCO_3$ for 10 days

($FeH_8N_2O_8S_2$) with Ferroin ($C_{36}H_{24}FeN_{62}{}^{+}$) as indicator. The color change at end-point for a positive result is from brown to purple. TAPPI procedure for alpha cellulose extraction was done and the bamboo fiber samples from bolo (***Gigantochloa bolo***) culm and running bamboo (***Phyllostachys aurea***) branches gave a purple end-point after the titration process. This signifies the presence of α-cellulose from the extracted bamboo fibers. This means that the extracted bamboo fibers contain the undegraded form of cellulose, α-cellulose.

Table 7.2 The measured external fiber diameter depicting fiber fineness from the extracted bolo (***Gigantochloa bolo***) culm and running bamboo (***Phyllostachys aurea***) branches

Sample 1	Pre-treatment	External fiber diameter (μm)
1—bamboo culm	1M NaOH, 6 days	78.75 ± 70
2—bamboo culm	8 wt.% $NaHCO_3$, 6 days	122.50 ± 40
3—bamboo culm	8 wt.% $NaHCO_3$, 10 days	127.50 ± 170
4—running bamboo	1M NaOH, 6 days	17.88 ± 14
5—running bamboo	8 wt.% $NaHCO_3$, 6 days	70.38 ± 84
6—running bamboo	8 wt.% $NaHCO_3$, 10 days	52.00 ± 87

7.4 Conclusion

This study shows that cellulosic fibers can be extracted from bolo (***Gigantochloa bolo***) culm and running bamboo (***Phyllostachys aurea***) branches grown in Baguio City, using $NaHCO_3$ pre-treatment and food-grade processing reagents. The Fourier Transform Infrared (FTIR) analysis revealed that the use of food grade reagents in the fiber extraction process was effective in removing surface impurities as shown in the general decrease in percentage absorbance (%A) of functional groups associated to hemicellulose and lignin for running bamboo (***Phyllostachys aurea***). This is also observed in the increasing crystallinity index (CI) calculated from x-ray diffractogram for six days $NaHCO_3$ pre-treatment and 10 days $NaHCO_3$ pre-treatment of bolo culms (***Gigantochloa bolo***). However, in terms of fiber diameter and fibrous structure it was seen that there is a need for longer $NaHCO_3$ pre-treatment time to achieve a finer structure and a more longitudinal feature for running bamboo (***Phyllostachys aurea***) fiber. The volumetric analysis using potassium dichromate ($K_2Cr_2O_4$) and ferrous ammonium sulfate ($FeH_8N_2O_8S_2$) in a reduction–oxidation reaction confirmed the presence of alpha(α) cellulose in the bamboo samples as evidenced by a purple end-point. While sodium hydroxide (NaOH) pre-treatment and the use of non-food grade reagents are effective in the extraction of a more crystalline and finer bamboo fibers, the use of food grade reagents such as sodium bicarbonate ($NaHCO_3$), acetic acid (CH_3COOH) in vinegar, and food grade hydrogen peroxide (H_2O_2) can also produce a comparable result when processing parameters are improved. It can be suggested that longer-pre-treatment time and processing time using food-grade reagents be explored to separate a more crystalline and cellulosic fiber from the bamboo fiber bundle. Bamboo samples should be chopped into smaller size before $NaHCO_3$ pre-treatment to allow lignin, a component of the fiber bundle depolymerize and solvate in the alkaline environment. Depolymerization occurs when bonds at β-O-4 (aryl-glyceryl-β-aryl ether bond) and 4-O-5 bond (diaryl ether bond) break. In the same way, smaller oligomers and xylose are also produced in a basic environment by hydrolysis of hemicelluloses. It is therefore concluded that the use of food grade reagents is possible in the extraction of cellulosic bamboo fibers from bolo culm and running bamboo to be used in food bio- packaging applications, edible bioplastics, textile and composite fabrication in industries—aeronautics, automotive, furniture.

Acknowledgements This work is funded by the Cordillera Studies Center (CSC), Individual Research Grant 2023.

References

1. N. Rehman, S. Alam, N. Ul Amin, I. Mian, H. Ullah, Ecofriendly isolation of cellulose from eucalyptus lanceolata: a novel approach. Int. J. Polym. Sci. **2018**(1), 1–7 (2018)

2. B. Bakri, A.E.E. Putra, A.A. Mochtar, I. Renreng, H. Arsyad, Sodium bicarbonate treatment on mechanical and morphological properties of coir fibres. Inter. J. Autom. Mech. Eng. **15**(3), 5562–5572 (2018)
3. P.P. Das, A. Manral, F. Ahmad, B. Sharma, V. Chaudhary, S. Gupta, P. Gupta, Environmentally sustainable chemical treatment of plant fibers for improved performance of polymeric composites. Polym. Compos. **43**(10), 7155–7169 (2022)
4. K.V. Parasuram, K.O. Reddy, M. Shukla, T. Marwala, Morphological, structural and thermal characterization of acetic acid modified and unmodified Napier grass fiber strands, in *2013 7th International Conference on Intelligent Systems and Control (ISCO)*. (IEEE Xplore, Canada, 2013), pp. 506–510
5. A. Belaadi, S. Amroune, M. Bourchak, Effect of eco-friendly chemical sodium bicarbonate treatment on the mechanical properties of flax fibres. Weibull statistics. Int. J. Adv. Manuf. Technol. **106**, 1753–1774 (2020)
6. A. Benkhelladi, H. Laouici, A. Bouchoucha, Tensile and flexural properties of polymer composites reinforced by flax, jute and sisal fibres. Int. J. Adv. Manuf. Technol. **108**, 895–916 (2020)
7. J.C.D. Santos, P.R. Oliveira, R.T.S. Freire, L.M.G. Vieira, J.C.C. Rubio, T.H. Panzera, The effects of sodium carbonate and bicarbonate treatments on sisal fibre composites. Mater. Res. **25**, e20210464 (2022)
8. G. Zheng, H. Ye, Y. Liang, X. Jin, C. Xia, W. Fan, Y. Xie, Y. Shi, J. Li, S. Ge, Pre-treatment of natural bamboo for use as a high-performance bio-composites via acetic acid ball milling technology. Constr. Build. Mater. **367**, 130350 (2023)
9. N.T. Phong, T. Fujii, B. Chuong, K. Okubo, Study on how to effectively extract bamboo fibers from raw bamboo and wastewater treatment. J. Mater. Sci. Res. **1**(1), 144 (2012)
10. P. Zakikhani, R. Zahari, M.T.H. Sultan, D.L. Majid, Extraction and preparation of bamboo fibre-reinforced composites. Mater. Des. **63**, 820–828 (2014)
11. S.M. Mortazavi, M.K. Moghadam, Introduction of a new vegetable fiber for textile application. J. Appl. Polym. Sci. **113**(5), 3307–3312 (2009)
12. L. Segal, J.J. Creely, A.E. Martin, C.M. Conrad, An empirical method for estimating the degree of crystallinity native cellulose using the x-ray diffractometer. Text. Res. J. **29**(10), 786–794 (1959)
13. L. Nayak, S.P. Mishra, Prospect of bamboo as a renewable textile fiber, historical overview, labeling, controversies and regulation. Fash. Text. **3**, 2 (2016)
14. T 203 cm-99 TENTATIVE STANDARD. Approved by the Chemical Properties Committee of the Process and Product Quality Division TAPPI. https://www.tappi.org/globalassets/t429-d-2-b-3-sarg.pdf (1999). Accessed 01 October 2024
15. L.Y. Mwaikambo, M.P. Ansell, Chemical modification of hemp, sisal, jute, and kapok fibers by alkalization. J. Appl. Polym. Sci. **84**, 2222–2234 (2002)
16. M.Y. Hashim, A.M. Amin, O.M.F. Marwah, M.H. Othman, M.R.M. Yunus, N.C. Huat, The effect of alkali treatment under various conditions on physical properties of Kenaf fiber. J. Phys. Conf. Ser. **914**, 012030 (2017)
17. M. Hu, C. Wang, C. Lu, N.I.S. Anuar, H.S.Y. Sheraz, M. Jing, Z. Chen, S. Zakaria, H. Zuo, Investigation on the classified extraction of the bamboo fiber and its properties. J. Nat. Fibers **17**(12), 1798–1808 (2020)

Chapter 8
Impact of pH Value and Utilization Rate Variation on Grinding Efficiency of Polycarboxylate-Based Grinding Aids

Veysel Kobya, Yahya Kaya, and Ali Mardani

Abstract In this study, the effect of pH variation of polycarboxylate-based (PCE) grinding aids (GAs) on clinker grinding efficiency, CO_2 emission, and environmental impact was investigated. For this purpose, the pH values of PCEs with constant chain lengths and molecular weights were varied as 4, 7, and 9. These PCE-based grinding aids (GAs) were applied at ratios of 0.05% and 0.1% of the combined weight of clinker and gypsum. The grinding process was conducted to achieve the target Blaine value of 3700 ± 100 cm^2/g. In total, six different cements were produced in addition to a control cement without PCE-based GAs. Grinding efficiency, CO_2 emission, and environmental impact were analyzed based on the grinding energy consumed during production. The study found that changes in utilization rate and pH value did not significantly impact the grinding performance of different types of PCE-based GAs. It was observed that PCE-based GAs improved grinding efficiency by 8–10% compared to the control cement without GAs.

Keywords PCE-based grinding aids · Grinding efficiency · Environmental impact

8.1 Introduction

Many factories aim to reduce energy consumption, emissions, and production costs. Optimizing energy use without compromising product quality is key to efficiency efforts, especially in stages like size reduction [1]. Grinding remains one of the most energy-intensive, inefficient processes in mineral processing [2]. Cement production, which consumes significant energy and emits greenhouse gases, focuses on optimizing clinker grinding, where two-thirds of the energy is spent [3]. Factors influencing energy consumption include mill type, speed, feed size, and grinding

V. Kobya (✉) · Y. Kaya · A. Mardani
Uludağ University, Görükle Campus, 16059 Nilüfer/Bursa, Turkey
e-mail: v.kobya@gmail.com

C. Areeprasert et al. (eds.), *Proceedings of the International Conference on Sustainable Energy and Environmental Technology for Circular Economy*, Springer Proceedings in Energy, https://doi.org/10.1007/978-981-96-2306-8_8

media [4, 5]. The use of Grinding Aids (GA) is a method to enhance energy efficiency and reduce emissions [6–8]. Recent research on polycarboxylate-ether-based GAs suggests similar performance to traditional GAs, enhancing workability and potential use if decomposition temperatures improve [9, 10]. Due to the structure of PCE, parameters like chain lengths, pH level, and the type of monomer used can be modified to achieve desired properties. Similarly, certain adjustable PCE parameters can influence milling performance. However, given the large number of variables, there is a significant gap in the literature concerning the effect of specific PCE properties, such as pH level, on grinding performance.

In this context, a new generation PCE-based PCE with 3 different pH values was synthesized. It aimed to evaluate the effects of the synthesized new-generation additives on grinding efficiency, CO_2 emissions, and overall environmental impact. GAs were used at 0.05 and 0.1% of the combined weight of clinker and gypsum. For this reason, grinding efficiency, CO_2 emissions, and environmental effects were investigated in cements with a Blaine fineness of 3700 ± 100 cm^2/g.

8.2 Material and Methods

In this study, three different PCE-based GAs were synthesized at different pH values and showed water-reducing properties. Table 8.1 shows some properties of the synthesized PCE-based GAs.

The reference cement mixture consisted of 96% clinker and 4% gypsum, following the TS EN 197-1 standard, and did not include any grinding aid (GA). Grinding was conducted using the ball mill, following the Bond standard procedure. For the cements containing PCE-based GAs, the grinding process used 96% clinker and 4% gypsum, with different PCE-based GAs added at two ratios (0.05% and 0.1%) of the total weight. In selecting these utilization rates, practical preferences and existing studies in the literature [6–8, 10] were taken into account. This resulted in the production of seven distinct types of CEM I 42.5 R cement, each achieving a Blaine fineness of 3700 ± 100 cm^2/g. When naming the cements, we considered the type and dosage of the GA used. For instance, cement containing 0.05% PCE4 was named PCE4-0.05, and cement without GA was labeled as the reference.

The physical and chemical properties of the clinker and gypsum, as provided by the manufacturer, are detailed in Table 8.2. The Bond Standard was used for preparing clinker and gypsum for grinding. Clinker and gypsum were passed through a 3.35 mm

Table 8.1 Some chemical properties of PCE-based GA

GA type	Main chain length (Da)	Side-chain length (Da)	Molecular weight (Da)	pH
PCE 4	4000	1000	19,000	4
PCE 7	4000	1000	19,000	7
PCE 9	4000	1000	19,000	9.1

sieve during this process. The ball distributions used in the grinding phase are given in detail in Table 8.3. The energy consumed during grinding is calculated using the Eq. 8.1.

$$Eg = (Tg * 1000) / (m * Tv) \tag{8.1}$$

where Eg represents the grinding energy (kWh/ton), Tg is the electricity recorded by the meter (kWh), m is the mass of the feed (clinker and gypsum) (kg), and Tv is the mill factor, set at 3 based on the manufacturer's specifications.

8.3 Results and Discussions

8.3.1 Grinding Efficiency

Table 8.4 presents the grinding times needed to achieve the target Blaine fineness value of 3700 ± 100 cm^2/g for the cements after grinding and the energy consumed during the grinding process.

As shown in Table 8.4, energy efficiency increased with the use of GA, regardless of their type or dosage. Considering the average dosages of Gas, PCE7 emerges as the highest-performing GA, as shown in Table 8.4. However, it is important to note that no significant difference in grinding efficiency was found between PCE7 and PCE4. In addition, considering the GA usage rate, the grinding efficiency was negatively affected by the increase in pH value. Some authors stated that the grinding performance is affected by the pH of the admixture [11]. It was also determined that reducing the pH of GA from 8 to 4 causes a higher Blaine fineness value at the same grinding time [12]. According to Vieira and Peres [13], grinding efficiency improved with an increase in GA pH from 7.3 to 10. Generally, grinding performance was found to enhance as pH diverges from the neutral level of 7 [14].

8.3.2 CO_2 Emissions and Environmental Impact

2021 Turkey produced 64.85 million tons of cement, according to the Turkish Cement Manufacturers Association [15]. The Ministry of Energy and Natural Resources of the Republic of Turkey reported an emission factor of 0.7258 tCO_2/MWh for 1 kWh of electrical energy on the national electricity network in the same year [15, 16]. This study utilized these data to assess the impact of GA utilization on the annual energy consumption and CO_2 emissions of the cement grinding process in Turkey. The findings are summarized in Table 8.5.

Table 8.5 shows that approximately 2,879,526 MWh of electrical energy was used in the clinker grinding process for cement production in Turkey in 2021. It has been

Table 8.2 Chemical contents of clinker and gypsum

Chemical composition		SiO_2	Al_2O_3	Fe_2O_3	CaO	MgO	SO_3	Na_2O	K_2O	Cl	C_3S	C_2S	C_3A	C_4AF	Free Lime	LoI
Content (%)	Clinker	21.1	5.58	3.26	66.7	1.11	0.5	0.51	0.68	0	62.7	13	9.28	9.91	1.45	0.3
	Gypsum	2.43	0.7	0.37	31.8	0.56	42	0.15	0.1							

Table 8.3 Ball mill conditions used in the grinding process

Ball distribution	Number of balls with a						
	90 mm	70 mm	60 mm	50 mm	40 mm	30 mm	20 mm
Number of balls used	2	2	10	15	30	75	75

Table 8.4 Grinding performances of GA

	Blaine value (cm^2/g)	Number of revolutions required for target Blaine	Grinding Time (min) for target Blaine	Energy Spent (kWh)	Relative energy efficiency (%)	Average relative energy efficiency (%)
Control	3800	7600	89,41	37,54		
PCE4-0.05	3620	6720	79,06	33,19	11,58	12.7
PCE4-0.01	3800	6550	77,06	32,35	13,82	
PCE7-0.05	3640	6620	77,88	32,70	12,89	13.03
PCE7-0.1	3680	6600	77,65	32,60	13,16	
PCE09-0.05	3700	7240	85,18	35,76	4,74	3.75
PCE9-0.1	3620	7390	86,94	36,50	2,76	

Table 8.5 The energy quantities and CO_2 emissions are computed based on the annual cement production in Turkey

CO_2 emissions and energy efficiency	Control	PCE4	PCE7	PCE9
The achieved energy savings (MWh)		307,408	317,136	98,578
He quantity of CO_2 emitted due to electricity consumption during grinding (tons)	1,767,046	1,543,929	1,536,869	1,695,498
The quantity of CO_2 emissions avoided (tons)	–	223,117	230,177	71,548

estimated that approximately 307,408 MWh of electricity will be saved by using PCE4. Similar savings in cements containing PCE7 and PCE 9 were calculated as 317,136 MWh and 98,578 MWh, respectively. According to Savaş [16], a household with two children in Turkey consumes approximately 2600 kWh of electrical energy annually. Therefore, using PCE4 during the clinker grinding process could save as much electricity as approximately 118,000 households use in a year. This saving is estimated to be 122,000 households for PCE7 and 38,000 households for PCE9 cements. Research shows that approximately 75% of electricity in Turkey is produced from fossil fuels that produce greenhouse gas emissions [15]. Assuming that an average tree captures and converts 12 kg of CO_2 per year during photosynthesis, the carbon footprint reduction due to trees can be calculated as follows: 1 tree captures 0.012 tons of CO_2 per year [16]. In this context, PCE7 was the contribution that resulted in the most significant reduction in the carbon footprint.

8.4 Conclusion

This study examined the energy efficiency and environmental impact of PCE-based Gas synthesized at various pH levels in cement production. Key findings include:

- GA usage improved grinding performance across all types and dosages.
- The highest performance was achieved with PCE-based GA at pH 7.
- Using GA at pH 7 can save about 320,000 MWh of energy and prevent 239,000 tons of CO_2 emissions.

Future research should explore the workability and rheology of systems using cements produced with different PCE-based Gas.

Acknowledgements The authors acknowledge support from Bursa Uludağ University Science and Technology Center (BAP) under grant FGA-2022-1151.

References

1. T. Zhang, J. Gao, J. Hu, Preparation of polymer-based cement grinding aid and their performance on grindability. Constr. Build. Mater. **75**, 163–168 (2015)
2. L.G. Austin, R.R. Klimpel, 4P.T. Luckie, *Process Engineering of Size Reduction Ball Milling*. (Society of Mining Engineers, AIME, New York, 1984)
3. K.G. Tsakalakis, G.A. Stamboltzis, Correlation of the Blaine value and the d80 size of the cement particle size distribution. ZKG Int. **61**(3), 60 (2008)
4. G. Danha, N. Hlabangana, E. Muzenda, Effect of ball and feed particle size distribution on the milling efficiency of a ball mill: an attainable region approach. S. Afr. J. Chem. Eng. **25**(1), 79–84 (2018)
5. T. Santosh, C. Eswaraiah, R.K. Soni, S. Kumar, Size reduction performance evaluation of HPGR/ball mill and HPGR/stirred mill for PGE bearing chromite ore. Adv. Powder Technol. **34**(1), 103907 (2023)
6. J.J. Assaad, S.E. Asseily, J. Harb, Effect of specific energy consumption on fineness of Portland cement incorporating amine or glycol-based grinding aids. Mater. Struct. **42**(8), 1077–1087 (2009)
7. J.J. Assaad, C.A. Issa, Effect of clinker grinding aids on flow of cement-based materials. Cem. Concr. Res. **63**, 1–11 (2014)
8. Y. Kaya, V. Kobya, A. Mardani, J.J. Assaad, Effect of modified triethanolamine on grinding efficiency and performance of cementitious materials. Talanta Open 100293 (2024)
9. H. Yang, J. Plank, Z. Sun, Investigation on the optimal chemical structure of methacrylate ester based polycarboxylate superplasticizers to be used as cement grinding aid under laboratory conditions: effect of anionicity, side chain length and dosage on grinding efficiency, mortar workability and strength development. Constr. Build. Mater. **224**, 1018–1025 (2019)
10. Z. Sun, H. Yang, L. Shui, Y. Liu, X. Yang, Y. Ji, Q. Luo, Preparation of polycarboxylate-based grinding aid and its influence on cement properties under laboratory condition. Constr. Build. Mater. **127**, 363–368 (2016)
11. H. El-Shall, P. Somasundaran, Physico-chemical aspects of grinding: a review of use of additives. Powder Technol. **38**(3), 275–293 (1984)
12. P. Halasyamani, S. Venkatachalam, R. Mallikarjunan, Influence of pH on kinetics of comminution of quartz. Ind. Eng. Chem. Process. Des. Dev. **7**(1), 79–83 (1968)

13. M.G. Vieira, A.E.C. Peres, Effect of rheology and dispersion degree on the regrinding of an iron ore concentrate. J. Market. Res. **2**(4), 332–339 (2013). https://doi.org/10.1016/j.jmrt.2013.07.002
14. D.W. Fuerstenau, Grinding aids. Kona Powder Part. J. **13**, 5–18 (1995)
15. Turkish Cement Manufacturers Association Cement statistics produced annually in Turkey, Homepage (2021), https://www.turkcimento.org.tr/tr/istatistikler/aylik-veriler. Accessed 2021
16. D.Ö. Savaş, Preparation of greenhouse gas inventory of Bursa Inegol district and greenhouse gas reduction plan within the scope of climate change action plan, Master's thesis, Bursa Uludağ University (2021)

Chapter 9
Hydrophobic Honeycomb-Like Tubular Biochar Derived from Corn Stalks as a Potential Absorbent for VOCs Removal Under Humid Conditions

Fanthagiro Rossi Stuart Majing, Yen San Chan, Yie Hua Tan, and Mohd Dinie Muhaimin Samsudin

Abstract This study focuses on the development of effective hydrophobic modified honeycomb-like tubular biochar (HT-B) with enhanced adsorption capacity under humid environment. HT-B was first synthesized from corn stalks (CS) and underwent surface modification through the impregnation of metal salt to produce CS/HT-B_{1M}. It was further modified through hydrophobic coating using polydimethysiloxane (PDMS). Characterization studies and VOCs adsorption studies were conducted at relative humidities of 50, 70, and 90%. It was found that the coated CS/HT-B_{1M} exhibits the highest adsorption capacity at coating ration of 1:25 with the modification temperature of 200 °C for an hour. As the relative humidity increases, the coated CS/HT-B_{1M} shows reduction in the adsorption capacity of 28.7% compared to the uncoated form. The findings showed that PDMS coated CS/HT-B_{1M} enhances the adsorption capabilities of biochar under humid conditions.

Keywords Adsorption · Honeycomb-like tubular biochar · Humidity · Hydrophobicity

F. R. Stuart Majing · Y. S. Chan (✉)
Department of Chemical & Energy Engineering, Faculty of Engineering & Science, Curtin University Malaysia, Miri, Sarawak, Malaysia
e-mail: chanyensan@curtin.edu.my

Y. H. Tan
Department of Chemical & Petroleum Engineering, Faculty of Engineering, Universiti Teknologi Brunei, Begawan, Brunei

M. D. Muhaimin Samsudin
School of Chemical & Energy Engineering, Faculty of Engineering, Universiti Teknologi Malaysia, Johor, Malaysia

C. Areeprasert et al. (eds.), *Proceedings of the International Conference on Sustainable Energy and Environmental Technology for Circular Economy*, Springer Proceedings in Energy, https://doi.org/10.1007/978-981-96-2306-8_9

9.1 Introduction

Various VOC abatement techniques have been developed to capture or remove the desired VOCs under different process conditions. Among, adsorption is one of the most economical options and has proven to be an efficient technology due to its wide range of flexibility and facile operation conditions [1]. Besides, desorption process enables VOCs to be recovered and recycle for further applications. Recently, adsorption techniques by using highly specific surface area adsorbents are attractive in the removal of VOCs due to their excellent adsorption capability. Porous materials including carbon-based materials, organic polymers, and oxygen-containing materials are commonly used adsorbents to enhance the adsorption of VOCs even at low concentrations [2]. Biochar has shown great potential as an adsorbent in environmental remediations due to its high specific surface area, adjustable pore structure, and high adsorption capability [3]. However, due to the natural properties of biomass, biochar synthesized from biomass wastes is mostly hydrophilic. Water vapor is unavoidable in the removal of VOCs due to its high vapor pressure at room temperature. In this case, VOCs and water molecules compete for the adsorption sites on biochar, leading to poor adsorption efficiency for the desired pollutant, which has been reported as low as 50% and remains unresolved [4]. This has become one of the major challenges in VOC adsorption technology.

In light of the limitations, multiple modifications, including structural and surface modifications are attractive approaches to enhancing the pore structure and surface properties of biochar. A unique honeycomb-like tubular structure has become the key to high adsorption efficiency. Its superior mechanical properties, honeycomb-like and tubular structure result in a high specific surface area, high adsorption capability, and good carrier transport properties [5]. Therefore, honeycomb-like tubular biochar (HT-B) played a significant role in improving the stability of the pore structure and increasing the specific surface area of the biochar at high modification temperatures. In our previous study, the HT-B was developed to investigate the pore structure development and its behavior in adsorbing VOCs under humid conditions. Remarkably, HT-B exhibited an impressive 3.45-fold increase in specific surface area while retaining a well-developed honeycomb-like tubular structure. However, at high relative humidity, the reduction of adsorption capacity was approximately 50%. This reduction highlights the significant impact of water molecules on the adsorption performance of HT-B.

Thus, surface modifications by introducing new surface functional groups are proven to be an effective method for altering the interaction between VOCs and biochar. Hence, in this study, the optimization of the hydrophobic coating conditions for the development of hydrophobic HT-B have been investigated. The hydrophobic HT-B were then used to investigate the performance of VOCs adsorption under different relative humidity. The structural properties, morphology, and hydrophobicity of hydrophobic HT-Bs were confirmed in this study. Moreover, to investigate the adsorption mechanisms involved and the performance of hydrophobic HT-Bs under different relative humidity conditions, The novelty of this study lies in its ability to fill the gap in the use of hydrophobic HT-B in gas pollutant adsorption applications, particularly under humid conditions. The study also provides insights into the optimal conditions for the preparation and modification of hydrophobic HT-B, which can be used to improve the adsorption performance of gas pollutants in industry.

9.2 Material and Methods

9.2.1 Materials

The PDMS (CAS No. 761036) were purchased from Sigma-Aldrich. The honeycomb-like tubular biochar (HT-B) derived from corn stalks was prepared in the previous study (Majing et al. 2023).

9.2.2 Surface Modification of HT-B

The optimization study utilized the one-factor-at-a-time (OFAT) approach, and the optimized parameters are presented in Table 9.1. The PDMS coating process was executed using the thermal evaporation decomposition method. This method entails the transformation of silicone molecules into short PDMS chains, which then form a conformal layer on the surface of CS/HT-B1M. Subsequently, these chains crosslink, creating a passivation layer that effectively eliminates the hydrophilic surface characteristics [6]. Three main optimization parameters for the PDMS coating are coating ratio, modification temperature, and reaction time.

Firstly, 25 g of PDMS was weighed and placed on the stainless-steel sieve pan, then the stainless-steel sieve with a pore measurement of 100–200 mesh was placed over it, as shown in Fig. 9.1. Next, 1 g of CS/HT-B1M was placed evenly on the sieve, and finally the sieve was sealed and heated in the furnace at modification temperature of 200 °C for 1 h [7]. Then, the coated CS/HT-B1M was cooled down at room temperature, collected and closed storage for the next applications. The yield was labelled as CS/HT-B1MPDMS. This procedure was repeated at different optimization conditions.

Table 9.1 Optimization parameters for the formation of CS/HT-B1MPDMS

No	Parameter	Unit	Range
1	Coating ratio (CS/HT-B_{1M}: PDMS)	g	1:15, 1:20, 1:25, 1:30
2	Modification temperature	°C	150, 200, 250, 300
3	Reaction time	h	0.5, 1, 1.5, 2

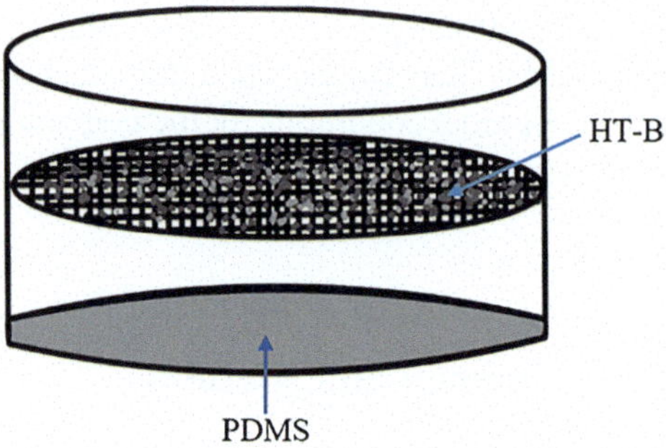

Fig. 9.1 PDMS-coated HT-B preparation

9.2.3 *Material Characterization*

The specific surface area, pore volume, and average pore diameter of the CS/BC, CS/HT-B1M, and CS/HT-B1MPDMS were analyzed by Brunauer-Emmett-Telle (BET). Their surface morphology and chemical compositions were studied by using a Scanning electron micrograph (SEM–EDS) Thermo Scientific Quattro S, which operated at 12.5–15 kV. Their surface functional groups were analyzed by the Attenuated total reflection infrared spectroscopy (ATR-IR) technique with transmission mode between 650- 4000 cm-1 using a Fourier transform infrared spectroscopy (FTIR) Cary 630.

9.2.4 *Adsorption Studies*

Figure 9.2 illustrates the setup for acetone adsorption, which comprises three distinct subsystems. The first subsystem involves gas generation, consisting of an air compressor serving as the carrier gas source. Acetone and water vapors are generated using a bubbler. Subsequently, the carrier gas conveys both acetone and water vapors to a mixer, then to the adsorption column subsystem. In the adsorption column, a U-shaped tube measuring 127 mm in height and 10 mm in inner diameter was immersed in a water bath to maintain the adsorption temperature at 25 °C.

A precise quantity of 1 g of CS/HT-B1MPDMS was placed at the bottom of the U-shaped tube, with both opening ends sealed using cotton wools. The flow rate of the carrier gas was monitored using a gas flowmeter set at 25 ml $min^{-}1$. The concentration of acetone vapors was maintained at 8 ml $g^{-}1$, and the relative humidity levels of 50, 70, and 90% were measured using a humidity hygrometer.

The final subsystem involves analysis via gas chromatography (GC-FID) and data storage using a computer. The inlet and outlet concentrations of acetone in the gas stream were continuously monitored by GC-FID, and the results was denoted as C_0 and C.

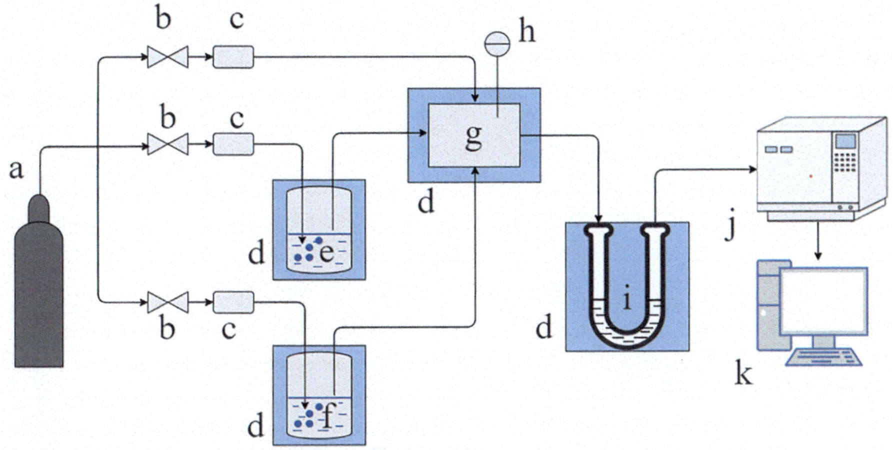

Fig. 9.2 Scheme of acetone adsorption: a. gas cylinder (compressed air); b. valve; c. mass flowmeter; d. water bath; e. acetone; f. deionized water; g. mixer; h. humidity hygrometer; i. adsorption column; j. gas chromatography; k. data collector [8]

9.3 Result and Discussion

9.3.1 Characterization of CS/HT-B$_{1M}$ and its Coated Samples

The previous study provided insights into the characterization of CS/HT-B1M [9]. The characterization of CS/HT-B1M has been discussed in our previous study. Subsequently, FTIR analysis was performed to study the changes of the surface functional groups of CS/HT-B1M after the surface modification with PDMS coating. Figure 9.3 shows the asymmetric and symmetric Si–O–Si stretching peaks observed at 779 and 1028 cm^{-1}, respectively, due to the introduction of PDMS on the surface of CS/HT-B1M. These peaks are only present on the coated biochar. Moreover, the CH_3 peak of the CS/HT-B1MPDMS at 1300 cm^{-1} is sharper than that CS/HT-B1M which attributed to the asymmetric stretching of Si–O–Si. This observation aligns with the increase in Si–O and CH3 groups resulting from the PDMS coating. These observations are further supported by the EDS mapping results of CS/HT-B1MPDMS as presented in Table 9.2. The results show that the atomic percent of Si (silicon) and O (oxygen) on the surface of CS/HT-B1MPDMS increased from 0% to 3.8% and from 3.1% to 6.7%, respectively. The increment of Si and O further confirms the successful application of the PDMS coating and its consequential impact on the surface chemistry of the CS/HT-B1MPDMS as illustrated in Fig. 9.4. Notably, the increased atomic ratio of Si and O for CS/HT-B1MPDMS is close to 1: 1.1, indicating the formation of an ultrathin PDMS film on the surface of CS/HT-B1M. Consistent with previous studies an increase in the atomic ratio approaching 1:1.5 and 1:1.2 is considered indicative of an ultrathin PDMS film [3, 4, 7]. Furthermore, a slight increase in the atomic ratio can be attributed to the incomplete oxidation of silicon

Fig. 9.3 FTIR spectra of CS/BC, CS/HT-B1M, and CS/HT-B1MPDMS

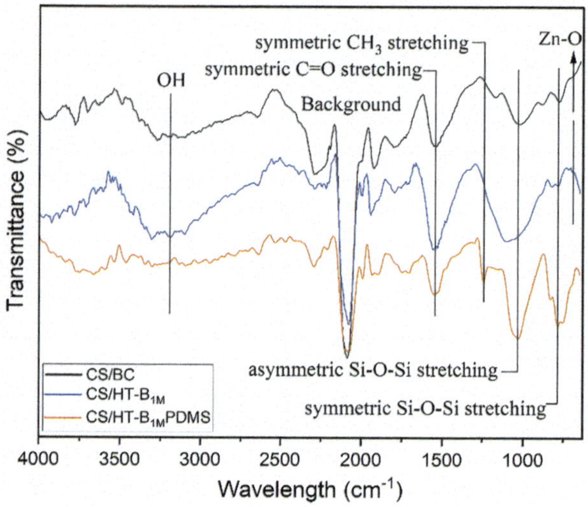

Table 9.2 The summary of the atomic percent of the HT-Bs and CS/HT-B1MPDMS. Sample EDS Analysis (wt %)

Sample	EDS Analysis (wt %)					
	C	O	Zn	Cl	Si	S
CS/HT-B_{1M}	92.7	3.1	0.3	1.6	–	0.1
CS/HT-B_{1M}PDMS	88.9	6.7	–	0.5	3.8	–

dioxide (SiO2). These investigations further suggest that the atomic percentage of Si is directly correlated with the thickness of the PDMS layer deposited on the surface of carbon-based materials. Hence, a higher atomic percentage of Si signifies a thicker PDMS film coating the surface.

9.3.2 Adsorption Capacity

Figure 9.5 shows the acetone adsorption capacity of CS/HT-B_{1M} and CS/HT-B_{1M}PDMS at varying relative humidity (RH) of 50, 70, and 90%. The results indicate a reduction in acetone adsorption capacity for both adsorbents as RH increased from 50 to 90%. This suggests that the presence of water molecules has a significant impact on acetone adsorption. The acetone adsorption capacity of CS/HT-B_{1M} decreased by 24.3 and 49.8% when RH increased from 50 to 70% and 90%, respectively. This can be attributed to the increased amount of water clusters, which block and compete with acetone for the adsorption active sites. The results of water contact angle indicated that CS/HT-B_{1M} has a hydrophilic surface, which means water molecules occupy the

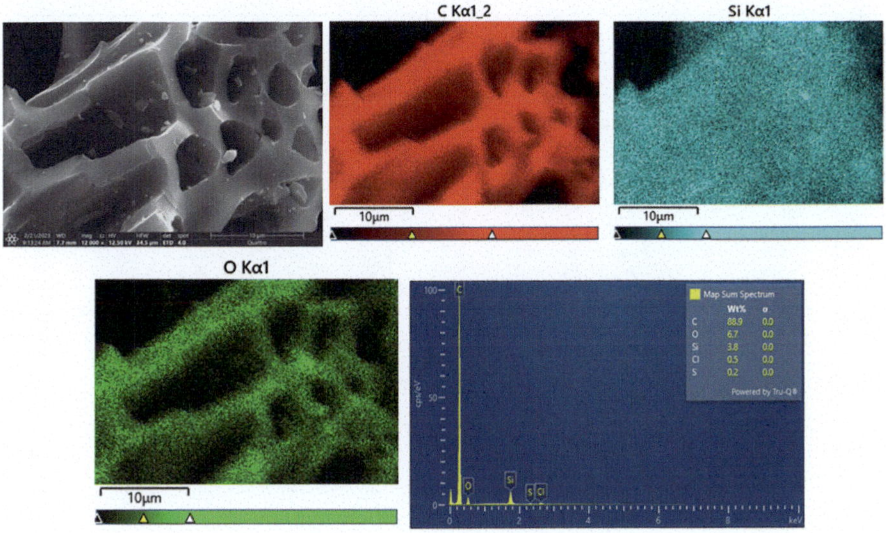

Fig. 9.4 EDS mapping images and spectra of CS/HT-B1MPDMS

adsorption active sites more easily than acetone, leading to poor adsorption performance at high RH, as described in our previous study [9] In contrast, the acetone adsorption capacity of CS/HT-B_{1M}PDMS decreased only by 15.1 and 28.7% when RH increased from 50 to 70% and 90%, respectively, indicating that the presence of water molecules has a lesser effect on acetone adsorption at high RH. Furthermore, CS/HT-B_{1M}PDMS exhibited a hydrophobic surface with a water contact angle of 132.8°. The hydrophobicity of CS/HT-B_{1M}PDMS reduced the affinity of water molecules and increased the selectivity of acetone adsorption at high RH. As a result, CS/HT-B_{1M}PDMS showed excellent acetone adsorption performance at high RH conditions.

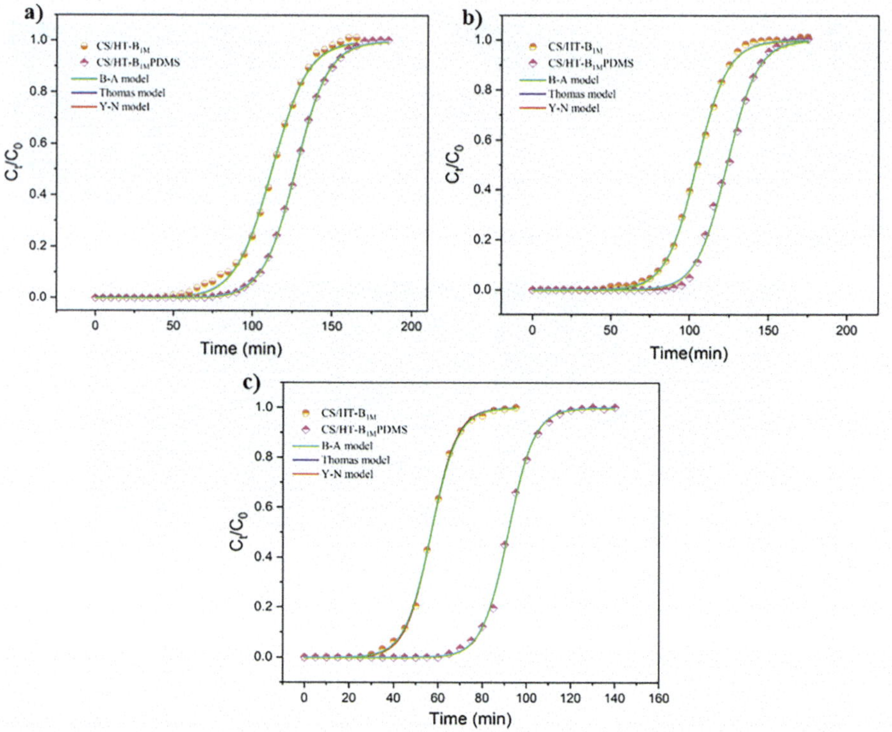

Fig. 9.5 Breakthrough curves of acetone adsorption onto CS/HT-B_{1M} and CS/HT-B_{1M}PDMS at **a** 50 RH% **b** 70 RH%, and **c** 90 RH

9.4 Conclusion

In summary, this study achieved the successful synthesis of hydrophobic honeycomb-like tubular biochar (hydrophobic HT-B) derived from corn stalks (CS). The process involved the hydrophobic coating of CS/HT-B_{1M} with polydimethylsiloxane (PDMS) through thermal evaporation, introducing siloxane (Si–O-Si) surface functional groups on the biochar. Several crucial parameters, including the coating ratio, modification temperature, and reaction time, were optimized to ensure the deposition of an adequate amount of PDMS onto the CS/HT-B_{1M} surface. The CS/HT-B_{1M}PDMS coated at a coating ratio of 1:25 (CS/HT-B_{1M}: PDMS), modification temperature of 200 °C, and reaction time of 1 h (CS/HT-B_{1M}PDMS). The coated CS/HT-B_{1M}PDMS also showed enhanced adsorption capacity of VOCs under the humid condition. Further research is needed to enhance the adsorption capacity at high humid condition > 80%.

Acknowledgements The authors would like to acknowledge the funding provided by Fundamental Research Grant Scheme (FRGS/1/2020/TK0/CURTIN/02/7)

References

1. K. Vikrant, K.H. Kim, W. Peng, S. Ge, Y.S. Ok, Adsorption performance of standard biochar materials against volatile organic compounds in air: a case study using benzene and methyl ethyl ketone. Chem. Eng. J. **387**, 123943 (2020)
2. R. Chen, Z. Yao, N. Han, X. Ma, L. Li, S. Liu, S. Wang, Insights into the adsorption of VOCs on a Cobalt-Adeninate Metal–Organic Framework (Bio-MOF-11). ACS Omega. **5**(25), 15402–15408 (2020)
3. H. Rajabi, M.H. Mosleh, T. Prakoso, N. Ghaemi, P. Mandal, A. Lea-Langton, M. Sedighi, Competitive adsorption of multicomponent volatile organic compounds on biochar. Chemosphere **283**, 131288 (2021)
4. L. Jia, J. Shi, C. Long, F. Lian, B. Xing, VOCs adsorption on activated carbon with initial water vapor contents: adsorption mechanism and modified characteristic curves. Sci. Total Environ. **731**, 139184 (2020)
5. X. Xu, Y. Guo, R. Shi, H. Chen, Y. Du, B. Liu, L. Li, Natural honeycomb-like structure cork carbon with hierarchical micro-mesopores and N-containing functional groups for VOCs adsorption. Appl. Surf. Sci. **565**, 150550 (2021)
6. Z. Gao, G. Song, X. Zhang, Q. Li, S. Yang, T. Wang, Y. Fu, A facile PDMS coating approach to room-temperature gas sensors with high humidity resistance and long-term stability. Sens. Actuators, B Chem. **325**, 128810 (2020)
7. Y. Zhang, C. Wei, B. Yan, Emission characteristics and associated health risk assessment of volatile organic compounds from a typical coking wastewater treatment plant. Sci. Total Environ. **693**, 133417 (2019)
8. T. Yin, X. Meng, L. Jin, C. Yang, N. Liu, L. Shi, Prepared hydrophobic Y zeolite for adsorbing toluene in humid environment. Microporous Mesoporous Mater. **305**, 110327 (2020)
9. F. R. Stuart Majing, Y. S. Chan, I. S. Tan, Y. H. Tan, M. D. Muhaimin Samsudin, Characterization of valorized pinewood sawdust to engineered activated biochar check for updates. Emerging technologies for future sustainability. In: Proceedings of the 2nd International Conference on Biomass Utilization and Sustainable Energy, pp 405–413. Springer Nature, Malaysia (2023)

Part III
Renewable Energy Technology

Chapter 10
The Effect of Heat Input and Heat Rejection on a Horizontal Gamma-Type Stirling Engine Performance

Chen Surapantanakorn, Chootrakul Siripaiboon, Kreetha Somkeattikul, Chaiyaporn Silawatchananai, Chinnathan Areeprasert, and Hang-Suin Yang

Abstract This study investigates the performance of a horizontal Gamma-type Stirling engine using air as the working fluid. The engine, which utilizes a slider-crank mechanism with two power pistons, was subjected to various heat inputs provided by torch burners, an electric band heater, and syngas from biomethanol production. The heat rejection was managed through both air and water cooling to compare their effectiveness. The primary objective was to measure the angular velocity and power output of the engine during its startup phase. The results indicated that the maximum angular velocity achieved was 225.9 rpm, with an average power output of 22.83 watts at a heater temperature of 445–455 °C. Water cooling system enhanced angular velocity and power output of the Stirling engine in all tests. However, results showed that water cooling system marginally improved the power output compared to the air cooling, the difference was not significant.

Keywords Stirling engine · Gamma-type · Torch burner · Water cooling · Syngas

---

C. Surapantanakorn · C. Siripaiboon · C. Areeprasert (✉)
Department of Mechanical Engineering, Faculty of Engineering, Kasetsart University, Bangkok 10900, Thailand
e-mail: fengcta@ku.ac.th

C. Siripaiboon
Department of Industrial Technology, School of Science and Technology, Sukhothai Thammathirat Open University, Nonthaburi 11120, Thailand

K. Somkeattikul
Department of Automotive Manufacturing Engineering, Faculty of Engineering and Technology, Panyapiwat Institute of Management, Nonthaburi 11120, Thailand

C. Silawatchananai
Department of Teacher Training in Mechanical Engineering, Faculty of Technical Education, King Mongkut's University of Technology North Bangkok, Bangkok 10800, Thailand

H.-S. Yang
Department of Mechanical Engineering, Faculty of Engineering, National Chung Cheng University, Chiayi 621301, Taiwan

C. Areeprasert et al. (eds.), *Proceedings of the International Conference on Sustainable Energy and Environmental Technology for Circular Economy*, Springer Proceedings in Energy, https://doi.org/10.1007/978-981-96-2306-8_10

10.1 Introduction

Stirling engines have been developed for many years since 1816s invented by Robert Stirling [1]. These engines are unique in their operation as external combustion engines with a closed thermodynamic cycle, be able to operate with various heat sources [2] and converts thermal energy to mechanical energy. In theory, the ideal regenerative Stirling cycle with a good regeneration is equal Carnot cycle that consists of two isothermal and isochoric processes. The Stirling engine divided into 3 types: Beta-type, Alpha-type and Gamma-type as written by Thombare and Verma [3].

The Gamma-type Stirling engine is known for its ability to operate under low-temperature differences. It's a simple mechanical design and a long lifespan, making it an attractive option in various applications. Although the power output produced by the Gamma-type engine is the lowest among Stirling engine types, its overall performance largely depends on the efficiency of its mechanism, as noted by Cheng and Yang [4]. Wen-Lin Chen et al. [5] presented a design for another Gamma-type Stirling engine, featuring twin power pistons with a similar construction aimed at the development and optimization of the Stirling engine. It was found that regeneration effectiveness affects efficiency, while the initial pressurization and crank radius of the power piston impact engine power. On the other hand, increasing the angular velocity of the engine also affects the engine power. Jerald Joseph et al. [6] showed the testing and manufacturing aspects of small-scale gamma type Stirling engine using lamp frame as heat input with CAD software and CNC machine. A temperature difference of about 180–200 °C was observed between the heat input and heat rejection. After modification and proper lubrication, the engine operated at a working speed of 2600 rpm.

This study focuses on the design, fabrication, and performance analysis of a horizontal Gamma-type Stirling engine using air as working fluid, which utilizes a slider-crank mechanism with two power pistons. Heat input was provided using a torch burner, heater, and syngas. The power output and angular velocity were analyzed based on the effects of heat input and heat rejection of the Gamma-type Stirling engine.

10.2 Material and Methods

10.2.1 Parametric Study

The size of the whole Gamma-type Stirling engine was 513 × 700 mm including displacer piston, power piston, crankshaft and flywheel. The gap between housing and displacer was 1 mm and the thickness of housing was 1.5 mm to cover the pressurization of air. The material of displacer and power pistons were Aluminum 6061 and Stainless Steel 304. For the power piston, a twin power piston was used to increase power from the system and dry seal was applied between the housing and

Table 10.1 Parameters and materials of Stirling engine's main parts

Part	Parameter of diameter (mm)	Material
Displacer piston	68	Aluminum 6061
Power piston	70	Aluminum 6061
Fly wheel	100	Stainless steel 304
Crankshaft	15	Stainless steel 304

piston. The crank radius of power piston and displacer pistons was 35 mm. Phase angle was 90 degrees. The main parameters and materials used for each part are presented in Table 10.1.

10.2.2 Configuration of Stirling Engine

The Stirling engine was designed and fabricated with a slider-crank horizontal mechanism using the gamma type which included one displacer and twin pistons in separate cylinders, along with a crankshaft and two flywheels. The configuration of Stirling engine and its auxiliary system is illustrated in Fig. 10.1. In the process, heat input and heat rejection were used as thermal energy for the expansion and compression of air as the working fluid inside the displacer cylinder. The displacer piston converted thermal energy into mechanical energy. The mechanical energy from the displacer piston was transferred to the power piston, driving the entire Stirling system. The crankshaft obtained force from the power piston and converted the reciprocating motion into rotary motion. The flywheel contained mechanical energy to balance the engine, ensuring continued good performance. The Stirling engine system consists of the Stirling engine, heat source, heat rejection source, torque and RPM measurement, data logger, and computer.

10.2.3 Test Procedure

The experiment for the Stirling engine aimed to find the torque and angular velocity values using a torque transducer model TCR-2Kmf and a load cell model with a powder brake, model POB-01s2, controlled by a tension meter. This setup converts a voltage difference into the resistance force of the powder brake. The power output came from the mechanical energy of the flywheel. In experiment, three sources of heat input were utilized. The two types of heat rejections were tested: ambient and water cooling, for comparison. For the engine power output, it was calculated by Eq. 10.1, where P_S (Watt) is the power output of the Stirling engine, T (N•m) is torque, and ω (rpm) is angular velocity.

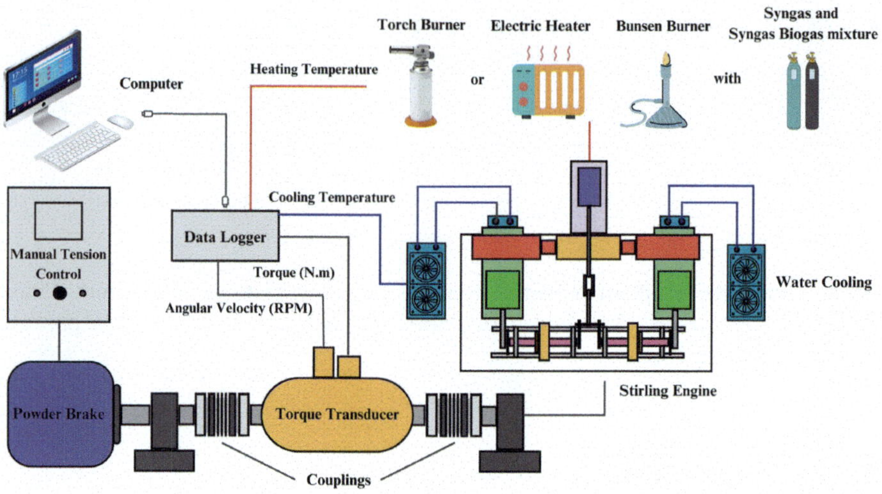

Fig. 10.1 Configuration of Stirling engine

$$P_S = T\omega 2\pi/60 \tag{10.1}$$

10.2.4 Heat Input and Heat Rejection

The heat sources tested in this work included a butane torch burner, a Bunsen burner using syngas from biomethanol production as fuel, and a electric band heater set at 350–500 °C. The energy input when using electric band heater was calculated by measuring the heating duration and input current. Work input of heater, torch and Bunsen burner were calculated by Eq. 10.2 and 10.3 respectively, where W_{in} (J) is work input, I (Ampere) is electric current, V (Voltage) is potential difference in the circuit, m (m^3/s) is volumetric flow rate of syngas, and LHV (MJ/m^3) is lower heating value of the syngas.

$$W_{in} = IV \int dt \tag{10.2}$$

$$W_{in} = m\,LHV \int dt \tag{10.3}$$

10.2.5 Slider Crank Mechanism

A Stirling engine that used a slider-crank mechanism to convert rotary motion to reciprocating motion includes a crank, connecting rod, slider, and frame, as shown in a horizontal gamma-type Stirling engine. In this study, the engine operated using simple harmonic motion and piston motion equations, as given by Eqs. 10.4 and 10.5, respectively, where x(m) is the distance of the end of the connecting rod from the crank axle, l(m) is the length of the connecting rod, r(m) is the length of the crank, and α (degree) is the angle of the crank measured from top dead center.

$$x = r\cos\alpha + l \tag{10.4}$$

$$x = r\cos\alpha + \sqrt{l^2 - r^2\sin^2\alpha} \tag{10.5}$$

10.3 Results and Discussion

10.3.1 Effect of Different Heat Sources on Stirling Engine Performance

Torch burner system utilized two burners to provide thermal energy to system. Temperature is around 390–450 °C measured by the infrared thermometer. With water cooling system, the angular velocity and power output of the engine reached 200 rpm and 18.4 watts, respectively. Without water cooling (ambient air cooling system), the angular velocity decreased to 191 rpm, and the power output dropped to 16.28 watts. This showed the effect of heat rejection between water cooling and ambient air cooling system. With more efficient heat rejection mechanism, the engine showed higher angular velocity and power output.

Electric band heater was utilized and the target heating temperature was 350–500 °C. With water cooling system, the angular velocity and power output were 213.97 rpm and 20.69 watts at 345–355 °C, 217.72 rpm and 20.92 watts at 395–405 °C, and 225.9 rpm and 22.83 watts at 445–455 °C, respectively. When the ambient air cooling system was used, the angular velocity and power output dropped to 186.96 rpm and 17.65 watts at 345–355 °C, 203.77 rpm and 19.63 watts at 395–405 °C, and 216.1 rpm and 21.58 watts at 445–455 °C.

For syngas combustion as heat sources, two types of syngas were utilized: syngas from biomethanol production and its mixture with biomass. The compositions are presented in Table 10.2. When applied water cooling system, the angular velocity and power output were 187.06 rpm and 17.3 watts, respectively. Without water cooling system, the angular velocity slightly decreased to 186.22 rpm and the power output dropped to 16.18 watts. For a syngas–biogas mixture system with water cooling

Table 10.2 Composition of two types of syngas

Types of syngas	Composition
Syngas	N_2 8% + CO 14% + CH_4 17% + H_2 18% + CO_2 balance
Syngas–biogas mixture	N_2 5% + CO 10% + CH_4 28% + H_2 13% + CO_2 balance

mechanism, the angular velocity was 177.1 rpm with a power of 16.21 watts. Without water cooling system, the angular velocity decreased to 172 rpm and the power was 15.79 watts. Again, this showed that the water cooling system affected the angular velocity and engine power.

10.3.2 Comparison Between Water Cooling and Air Cooling systems in Different Heat Input Conditions

From the experiment, the implementation of water cooling enhanced the power output of the Stirling engine across three heat sources. The highest power output was achieved when using an electric band heater at a temperature range of 445–455 °C. This was followed by the torch burner, with syngas combustion yielding the lowest power output. However, adding water cooling system may increase equipment cost and the improvement of the performance was clear but not significant. Therefore, further investigation is necessary. The comparison of the average power output of water cooling and ambient air cooling systems is presented in Fig. 10.2.

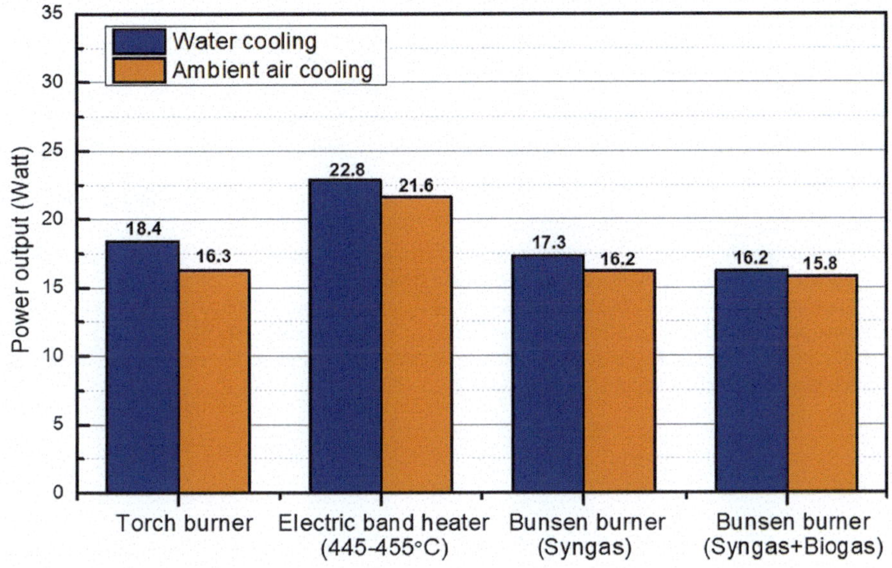

Fig. 10.2 Comparison of the average power output between water cooling and ambient air cooling

10.4 Conclusion

This study showed that Stirling engine was able to operate with various heat sources even high and low heating value. Conclusion can be made as follows:

1) The most of the components for gamma-type Stirling engine were fabricated using a CNC machine, employing materials such as Aluminum 6061, Stainless Steel 304, and brass. This choice of materials and machining method was aimed at enhancing the precision of alignment, reducing weight and vibration as well as minimizing friction between the interacting parts.
2) The horizontal Gamma-type Stirling engine achieved its highest angular velocity of 225.9 rpm and maximum power output of 22.83 watts when using a band heater operating at 445–455 °C with water cooling system.
3) Even though the values of power output with water cooling system are better than those without the water cooling system, the difference is not significant. It is also possible to reduce the work input by opting not to use water cooling system.
4) Accurately measuring the thermal efficiency of a torch and a Bunsen burner is challenging due to significant energy losses. However, for a heater operating mode, the efficiency was around 6–7%.

Acknowledgements This research received funding from the National Science, Research and Innovation Fund (NSRF) via the Program Management Unit for Human Resources & Institutional Development, Research and Innovation [grant number B40G660043].

References

1. M.Z. Getle, F. Lanzetta, B.T. Admassu, A.A. Hassen, Reversed regenerative Stirling cycle machine for refrigeration application: a review. Int. J. Refrig. **118**, 173–187 (2020)
2. C.-H. Cheng, H.-S. Yang, Stirling engine technology and its application on solar power generation, in *I-DAD* (Springer India, 2024), pp. 3–9
3. D.G. Thombare, S.K. Verma, Technological development in the Stirling cycle engines. Renew. Sustain. Energy Rev. **12**, 1–38 (2008)
4. C.-H. Cheng, H.-S. Yang, Optimization of geometrical parameters for Stirling engines based on theoretical analysis. Appl. Energy **92**, 395–405 (2012)
5. W.L. Chen, K.L. Wong, L.W. Po, A numerical analysis on the performance of a pressurized twin power piston gamma-type Stirling engine. Energy Convers. Manag. **62**, 84–92 (2012)
6. J. Joseph, E.M. Louis, B. Thomas, K. Anurag, V. Sankar, T.T. Pullan, Fabrication and testing of a gamma type Stirling engine. Mater. Today Proc. (2020)

Chapter 11
Study on the Degradation of H_2/O_2 PEMFC with a Dead-End Gas Management During Long-Term Operation

Panida Kanha, Werrachai Chaiworapuek, and Yong-Song Chen

Abstract The increasing environmental pollution problem has been attributed to fossil fuels, prompting a high demand for effective replacements. Proton exchange membrane fuel cells (PEMFCs) are a crucial technology for producing electric power in many applications that utilize hydrogen to generate high-efficiency and low-emission electricity. To improve the utilization of supplied gases, dead-end operations have been employed both on the cathode and anode sides of a fuel cell system; however, in dead-end operations, impurities may accumulate and cause the catalyst layer to degrade. In this study, both the anode and cathode of an H_2/O_2 PEMFC are operated in a dead-end mode. The cell is operated at 800 mA/cm^2 and 70 °C under various cathode gas humidifying conditions. The anode is purged at selected current integral values of 5000 As and 10,000 As, whereas the cathode is purged at 1400 As. The polarization curve, impedance, and morphology of the catalyst layer are investigated after long-term operation (24 and 48 h). The results indicate that the performance of the single cell deteriorates significantly under dry cathode conditions compared to 50%RHc. The minimum degradation rates observed with humidify on the cathode side were 0.059 and 0.076 V/h, respectively, with an anode current integral of 5000 As and 10,000 As. Furthermore, the thickness of the cathode catalyst layer decreased by 50% after 48 h of operation.

Keywords Proton exchange membrane fuel cell · Dead-ended · Purge · Degradation · Current integral

P. Kanha (✉) · Y.-S. Chen
Department of Mechanical Engineering, National Chung Cheng University, Chiayi 621301, Taiwan
e-mail: panida@alum.ccu.edu.tw

W. Chaiworapuek
Department of Mechanical Engineering, Kasetsart University, Bangkok 10900, Thailand

C. Areeprasert et al. (eds.), *Proceedings of the International Conference on Sustainable Energy and Environmental Technology for Circular Economy*, Springer Proceedings in Energy, https://doi.org/10.1007/978-981-96-2306-8_11

11.1 Introduction

Most of the world's energy needs are being met by fossil fuels, which have been harmful to the environment due to the human response since the Industrial Revolution, which has led to the extraction, transportation, processing, and combustion of such fuels. The main components of fossil fuels are carbon and hydrogen, and they also contain other elements, such as sulfur, as well as additives, including lead and alcohol, that are added during the refining process. When burned, these fuels release various gases (CO_X, SO_X, NO_X), soot, ash, tar droplets, and other organic compounds into the atmosphere, resulting in air pollution [1–3]. Hydrogen energy is a promising green fuel source that produces only water as a byproduct for electric power generation. This clean, sustainable energy source can solve many environmental problems. Proton exchange membrane fuel cells (PEMFCs), a crucial technology that uses hydrogen as fuel in many applications, offer a promising solution. These fuel cells utilize dead-end operations on both the cathode and anode sides to enhance the usage of supplied gases in a fuel cell system, thereby reducing the release of unused gas from the system [4, 5]. However, it is essential to consider the possibility of accumulating impurities and causing the catalyst layer's degradation when using this method. In a study, both the anode and cathode of an H_2/O_2 PEMFC were operated in a dead-end mode, analyzing the performance, the effects of various factors, and the long-term operation on the fuel cell performance. The research delves into the degradation of the catalyst through the utilization of electrochemical impedance spectroscopy (EIS) and scanning electron microscope (SEM) to observe the morphology of the catalyst layer.

11.2 Experimental

11.2.1 Experimental Setup

The experimental setup is illustrated in Fig. 11.1. A single PEMFC was used, which included a membrane electrode assembly (MEA) with an active area of 5×5 cm^2. The MEA was placed between carbon paper (28BC, SGL, Germany), graphite plates, gold-coated copper current collectors, and aluminum alloy end plates. The anode and cathode sides of the membrane (Nafion 212, Dupont, USA) were coated with 0.1 and 0.3 mg Pt/cm^2, respectively. The anode and cathode of the PEMF were both in a dead-end mode. The data acquisition module (NI9219, National Instruments, USA) was used to record cell voltage, temperature, current, and gas flow rate. At the cathode outlet, a normally-closed solenoid valve (Type 6126, Burkert, Germany) was meticulously controlled by a relay (KG1010D, KYTTO, Taiwan) and a digital output module (NI 9401, National Instruments, USA), ensuring the experiment's reliability.

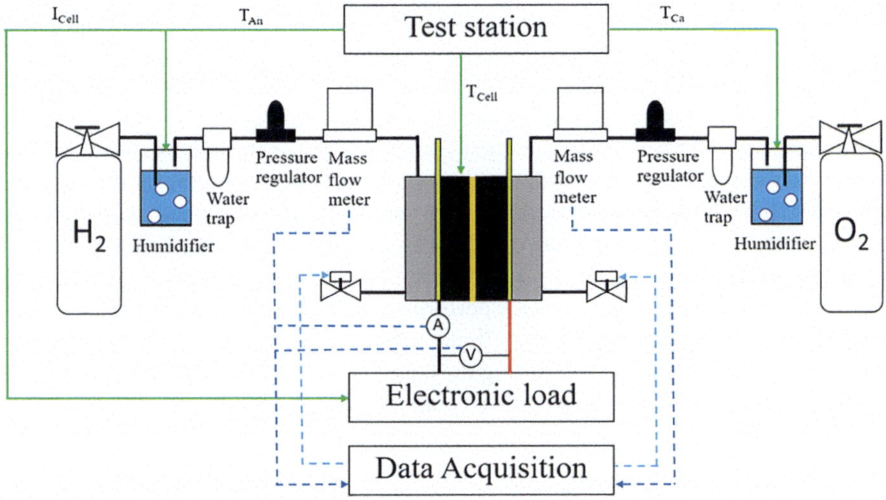

Fig. 11.1 A schematic of the experimental setup in this study

11.2.2 Experimental Procedure for Long-Term Operation

Specific conditions were maintained throughout the long-term performance experiments to accelerate the degradation rate. The cell temperatures were held at a consistent 70 °C, while the current density was kept at a precise 800 mA/cm^2. Additionally, to gain a comprehensive understanding of the impact of operational humidity on performance, the relative humidity cathode sides were carefully adjusted and varied with the anode fully humidified. The current integral is used as a purging strategy, with control at the cathode side maintained at 1400 As, while that of the anode side was maintained at 5,000 As and 10,000 As, which was investigated for up to 48 h. After long-term testing, the MEA cross-section morphology was analyzed using a scanning electron microscope (SU 8200, HITACHI, Japan).

11.3 Result and Discussion

11.3.1 Performance Degradation

The polarization curves at the beginning and end of each test day are used to observe the fuel cell stack's performance change, proving the effect of the purging strategy and humidification Fig. 2a shows that dry humidity at the cathode and longer purge duration by current integral accelerate performance degradation with degradation rates of 0.094 and 0.183 V/h with anode current integral of 5000 As and 10,000 As, respectively. The degradation rate of the cathode with 50% relative humidity is lower, as shown in Fig. 2b. Specifically, the degradation rates were 0.059 and 0.076 V/h, respectively, with an anode current integral of 5000 As and 10,000 As. The effect of humidity is also revealed in the EIS, as shown on the Nyquist plot in Fig. 11.3. Dry conditions have higher resistance dominated by membrane resistance than humidified conditions, which causes low performance, as shown on the polarization curve. Figure 11.3a shows that when the current integral for the cathode with 0%RHc is increased, ohmic resistancc incrcases. However, the chart transfer resistance of current integral 5000As decreases as water accumulates, aiding the cell in increasing oxygen reaction. An increase in chart transfer resistance over time of approximately 0.5 Ωcm^2 also causes a performance drop. However, in the humidified cathode 50%RHc in Fig. 11.3b, the ohmic resistance increases very little from the initial value of a fresh single cell during the purge in 5000As and 10000As. The chart transfer also increases that, causing performance to drop slightly over time.

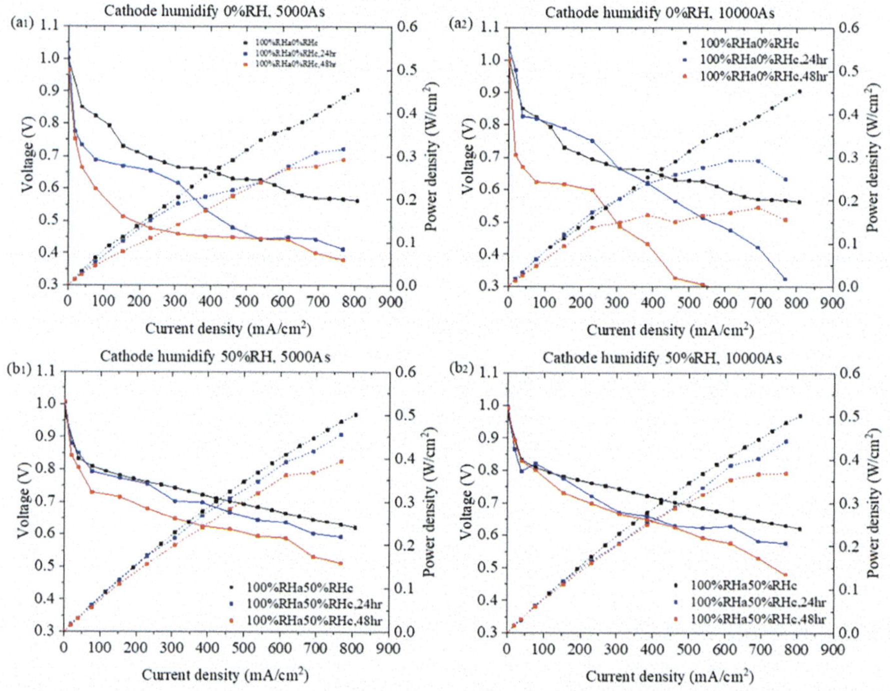

Fig. 11.2 The polarization curve of the single cells after the long-term test (1) Anode current integral 5000As; (2) Anode current integral 10000As with cathode humidity **a** 0%RHc and **b** 50%RHc

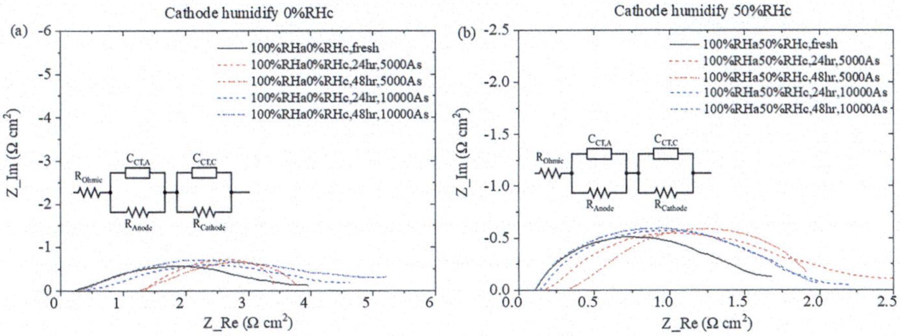

Fig. 11.3 The nyquist plot of the single cells after the long-term test with cathode humidity **a** 0%RHc and **b** 50%RHc

11.3.2 Cross-Section Morphology of Membrane Electrode Assembly (MEA)

During the dead-end mode operation, the loss of the catalyze layer may cause hydrogen starvation, resulting in carbon corrosion. According to Fig. 11.4a, the MEA in the fresh condition and the catalyst layer that coats the membrane are smooth and uniform. By coating with 0.1 and 0.3 mg Pt/cm^2, it is found that thickness is 12.16 and 25.5 μm on the anode and cathode side, respectively. In this case of fully humidifying at the anode and 50%RHa at the cathode, after the 48 h long-term test, the catalyzed layer decayed significantly. The thickness reduction was more than 50% of the original thickness. But in terms of integral, there is no significant difference. The loss of the catalyst layer is related to the platinum content used in the PEMFC because of its stability during the hydrogen and oxygen reactions at the anode and cathode, causing performance loss. However, the result of the current integral increase does not give the influential morphology decay of the catalyzed layer.

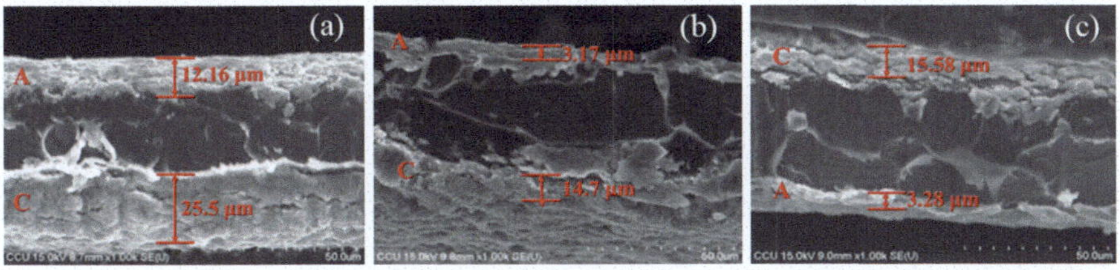

Fig. 11.4 The cross-section SEM on MEA investigates **a** Fresh **b** 5000As and **c** 10000As

11.4 Conclusion

Based on the results, it has been observed that the performance of a single cell deteriorates significantly under dry cathode conditions when compared to 50%RHc. The observed minimum degradation rates with humidified 50%RHc on the cathode side were 0.059 and 0.076 V/h, with an anode current integral of 5000 As and 10,000 As, respectively. The ohmic and chart transfer resistance by EIS also indicates that dry cathode results in worse performance and has a performance drop over time. Additionally, after 48 h of operation, the morphology of MEA lost the thickness of the cathode catalyst layer by 50%.

References

1. F. Barbir, T.N. Veziroğlu, H.J. Plass, Environmental damage due to fossil fuels use. Int. J. Hydrogen Energy. **15**(10), 739–749 (1990)
2. C.G. Solomon, R.N. Salas, D. Malina, C.A. Sacks, C.C. Hardin, E. Prewitt, T.H. Lee, E.J. Rubin, Fossil-fuel pollution and climate change - a new NEJM group series. N. Engl. J. Med. **386**(24), 2328–2329 (2022)
3. J. Yang, Y. Yu, T. Ma, C. Zhang, Q. Wang, Evolution of energy and metal demand driven by industrial revolutions and its trend analysis. Chin. J. Popul. Resour. Environ. **19**(3), 256–264 (2021)
4. J. C. Kurnia, A. P. Sasmito, T. Shamim, Advances in proton exchange membrane fuel cell with dead-end anode operation: a review. Appl. Energy. **252** (2019)
5. L. Fan, L. Xing, Z. Tu, S.H. Chan, A breakthrough hydrogen and oxygen utilization in a H_2-O_2 PEMFC stack with dead-ended anode and cathode. Energy Convers. Manag. **243** (2021)

Chapter 12
Experimental Study on the Effect of Electrode Configuration on the Performance of High Temperature Proton Exchange Membrane Fuel Cells

Pongsapak Treegosol and Yong-Song Chen

Abstract High–temperature proton exchange membrane fuel cells (HT–PEMFCs) are widely recognized as a feasible sustainable energy technology for usage in automotive and stationary applications. It has high CO tolerance, fast reaction kinetics, a high amount of reusable heat energy, and no humidification requirement due to the operating temperature higher than 100°C, thus the water at the inlet and the chemical reaction's products become the vapor phase, leading to simple water and thermal management. However, various factors influence the performance of high–temperature membrane electrode assemblies (MEAs) with phosphoric acid–doped PBI membrane, therefore determining the optimal operating parameters is one of the challenges for an HT–PEMFC. This study investigated the performance of HT–PEMFCs at various temperatures (140, 160, and 180°C), air cathode stoichiometric ratio (2, 2.5, and 3), acid doping levels (360 and 460%), and gas diffusion layers (GDL340, GDS with 5 and 20% PTFE). The catalyst layer was composed of Pt/C (46.6 wt. %), PVDF binder, and DMAc as a solvent. The Pt loading on the electrode was approximately set as 0.5 and 1 mg Pt·cm^{-2} for the anode and the cathode, respectively. The commercial PBI membrane was utilized in this study. The results indicated that increasing the operating temperature can enhance the performance of an HT–PEMFC. However, the cell performance also increased with increasing the air stoichiometric ratio at the cathode side and the acid doping level. Moreover, the MEA with GDS310 reached the maximum power density up to 0.410 and 0.396 W·cm^{-2}, which can be enhanced by approximately 31% and 27% under 5 and 20% PTFE, respectively.

Keywords High–temperature proton exchange membrane fuel cells · Polybenzimidazole · Acid doping level · Gas diffusion layers

P. Treegosol (✉) · Y.-S. Chen
Department of Mechanical Engineering, National Chung Cheng University, Chiayi 621301, Taiwan
e-mail: pongsapak@alum.ccu.edu.tw

C. Areeprasert et al. (eds.), *Proceedings of the International Conference on Sustainable Energy and Environmental Technology for Circular Economy*, Springer Proceedings in Energy, https://doi.org/10.1007/978-981-96-2306-8_12

12.1 Introduction

Nowadays, the rising consumption of electrical power leads to increased combustion of fossil fuels, which is a major contributor to environmental degradation due to carbon monoxide emissions and the depletion of energy resources. Alternative and renewable energy sources are vital for a sustainable energy cycle [1]. In recent years, fuel cells have attracted a lot of attention from researchers as an environmentally friendly, clean energy, zero emissions, and high-quality power scalability [2, 3], which is an electrochemical device that directly converts chemical energy from fuel into electrical energy. Moreover, HT–PEMFCs offer significant advantages over LT–PEMFCs, including improved gas impurities [4], faster kinetics reaction [5], and simplified water and heat management [6–8], making them highly promising for automotive, stationary, and portable applications [9]. Over the last decade, further studies have focused on HT–PEMFCs using phosphoric acid–doped Polybenzimidazole (PBI) membrane and investigated the operating parameters effect to enhance the performance and durability of fuel cells. Acid doping level (ADL) in the PBI membrane can improve the ionic transfer, leading to enhanced cell performance, while it can reduce the mechanical properties of the membrane, affecting the cell's durability [10]. Ohmic loss is the major factor that can affect the performance, which can be reduced by increasing temperature [11, 12]. In addition, various operating conditions must be considered to affect the performance, including the gases stoichiometric [12], the thickness of the gas diffusion layer [13], the effect of binder on the electrode and gas diffusion layer [13–16], the gases impurity [17]. Additionally, other researchers have investigated the bipolar plate and flow field development of HT–PEMFCs by numerical studies to find an optimal design structure and the optimal operating conditions [18, 19]. However, various factors influence the performance of high–temperature membrane electrode assemblies (MEAs) with phosphoric acid–doped PBI membrane, therefore determining the optimal operating parameters is one of the challenges for HT–PEMFCs. Thus, this study investigates the performance of HT–PEMFCs under various operating parameters, including the effect of operating temperatures (140, 160, and 180°C), air cathode stoichiometric ratio (2, 2.5, and 3), acid doping levels (360 and 460%), and gas diffusion layers (GDL340, GDS with 5 and 20% PTFE).

12.2 Experimental Methods

12.2.1 *Membrane Electrode Assemblies (MEAs) Preparation*

In this study, the catalyst inks consisted of a commercial catalyst with 46.6 wt. % Pt/C (Tanaka Kikinzoku Kogyo K.K., Japan), Polyvinylidene fluoride; PVDF binder, and Dimethylacetamide; DMAc as solvent. All contents were mixed using an ultrasonic bath and planetary. The catalyst inks were sprayed onto GDL 340 and

GDS 310 (CeTech Co., Ltd., Taiwan) using the ultrasonic spray coating technique. A commercial PBI membrane (hiPower Green Technology Co., Ltd., Taiwan) with an active area of 5×5 cm^2 was immersed in 85 wt. % H_3PO_4 at 30 °C until ADL reaches the desired level.

12.2.2 Fuel Cell Test

In a single stack fuel cell, the MEAs were assembled by sandwiching the PA–doped membrane between two GDEs. In all conditions, the Pt loading on the electrode was 0.5 and 1 mg Pt·cm^{-2} for the anode and the cathode side. The single–stack fuel cell consisted of the endplate, current collector, gasket, bipolar plate, and MEAs. Moreover, the fuel cell test station was carried out by HEPHAS Energy Co., Ltd. The fuel cell was activated under a constant voltage of 0.6 V at the temperature of 180 °C until the current became stable. Then, the steady–state polarization curves were obtained by measuring the voltage after 2 min at each current point. The experimental conditions were classified as shown in Table 12.1 to investigate the effect of operating temperatures, air cathode stoichiometric ratio, doping levels, and gas diffusion layers on the HT–PEMFCs performance.

Table 12.1 Experimental study conditions

Configurations	GDE		Cell temp (°C)	Stoi. ratio	ADL (%)
	Anode	Cathode			
1	GDL340	GDL340	140	1.2/2	360
2	GDL340	GDL340	160	1.2/2	360
3	GDL340	GDL340	180	1.2/2	360
4	GDL340	GDL340	160	1.2/2.5	360
5	GDL340	GDL340	160	1.2/3	360
6	GDL340	GDL340	160	1.2/2	460
7	GDL340	GDS310 (5% PTFE)	160	1.2/2	360
8	GDL340	GDS310 (20% PTFE)	160	1.2/2	360

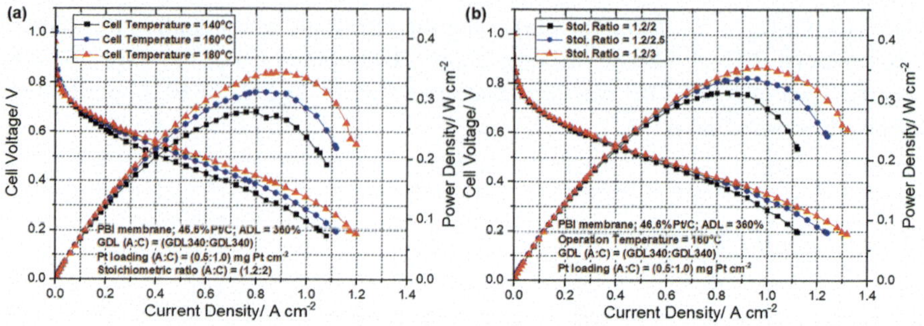

Fig. 12.1 Polarization curves under various **a** cell temperature and **b** air stoichiometric ratio

12.3 Results and Discussion

12.3.1 The Effect of the Operation Temperature

The performance curve under various operating temperatures (140, 160, and 180°C) was illustrated in Fig. 12.1a. The result showed that the cell performance tended to increase with increasing the operation temperature. When the cell voltage reached 0.6 V, the current density was 0.212, 0.256, and 0.300 A·cm^{-2} under the cell temperature of 140, 160, and 180°C, respectively. Moreover, the decreasing rate of the voltage in the ohmic losses region under low temperatures was significantly higher than at high temperatures due to the ohmic resistance tends to decrease with increasing cell temperature, leading to high performance under high operation temperatures. Furthermore, the maximum power density was 0.278, 0.312, and 0.345 W·cm^{-2} under the cell temperature of 140, 160, and 180°C, respectively.

12.3.2 The Effect of Air Stoichiometric Ratio

The performance curves under different air cathode stoichiometric ratios (2, 2.5, and 3) were observed in Fig. 12.1b. In this study, the cell was operated at 160°C, and commercial GDL340 was employed as GDEs on both anode and cathode sides. Figure 12.1b also showed that the cell performance tended to increase with increasing the air stoichiometric ratio, due to a high air stoichiometric ratio causing the high flowrate at the cathode side leading to the high chemical reaction inside the HT–PEMFCs when compared with the lower air stoichiometric. Moreover, when the cell voltage was observed at 0.6 V, the current densities were almost similar among these conditions, which were 0.256, 0.280, and 0.288 A·cm^{-2}, respectively. Meanwhile, at the cell voltage of 0.4 V, the current density significantly increased with increasing air stoichiometric ratio, which was 0.766, 0.832, and 0.876 A·cm^{-2}, respectively.

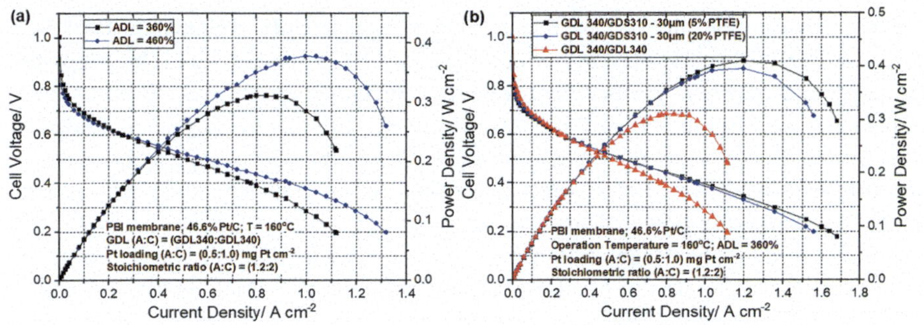

Fig. 12.2 Polarization curves under various **a** acid doping levels (ADL) and **b** gas diffusion electrode (GDE)

Furthermore, the peak power density (PPD) was 0.312, 0.337, and 0.354 W·cm^{-2} under the air stoichiometric ratio of 2, 2.5, and 3, respectively.

12.3.3 The Effect of the Acid Doping Level

Figure 12.2a showed the polarization curve under different acid doping levels – 360 and 460% on the MEAs, which were assembled by sandwiching the doped membrane between the GDL340 on both sides. These results also showed that at the low current density region (lower than 0.4 A·cm^{-2}), the cell voltage under various acid doping levels was almost similar due to the effect of the chemical reaction in these study didn't significant difference. On the contrary, at current densities higher than 0.4 A·cm^{-2}, the cell voltage at high ADL was significantly higher than the MEAs with lower ADL. Due to the ADL increases with decreasing the ohmic resistance, leading to high performance under high acid doping levels. According to the figure, when the cell voltage reached 0.4 V, the current density was 0.776 and 0.936 A·cm^{-2} under the acid doping level of 360 and 460%, respectively. Moreover, the maximum power density was 0.312 and 0.378 W·cm^{-2} under the acid doping level of 360 and 460%, respectively.

12.3.4 The Effect of the Gas Diffusion Layer with Different PTFE Content

The polarization curves under various compositions of cathode gas diffusion electrode (GDE) – GDL340, GDS310 (5% PTFE), and GDS310 (20% PTFE) were observed in Fig. 12.2b. In the study, the GDL340 was used as GDE at the anode side for all case studies, and the H_2 and air were supplied at the stoichiometric ratio of 1.2 and 2, respectively. The results found that when the GDS310 was used as GDE on the cathode side, the cell performance was significantly higher than GDL340. Because the MPL on GDS310 was composed of PTFE, which had a hydrophobic property and high gas permeability, resulting in the case of GDS310 with PTFE had higher performance than GDL340. Moreover, the cell performance tended to decrease with increasing the content of PTFE. When the cell voltage reached 0.4 V, the current density was 0.766, 0.960, and 0.948 A·cm^{-2} under three cases. Furthermore, the maximum power density under three cases was 0.312, 0.410, and 0.396 W·cm^{-2}, respectively. Meanwhile, when compared with GDL340, the performance in the case of GDS310 can be enhanced by approximately 31% and 27% under 5 and 20% PTFE, respectively.

12.4 Conclusion

In this study, the HT–PEMFCs performance under various operation temperatures, the air stoichiometric ratios, the gas diffusion layer with different PTFE contents, and the acid doping levels were investigated by polarization curves. As a result, the cell performance can be enhanced by increasing the operating temperature and the acid doping level (ADL) because of decreasing the ohmic resistance and, consequently high performance. However, the cell performance of HT–PEMFCs also increased with increasing the air stoichiometric ratio. Due to higher chemical reactions inside the cell, leading to higher performance. In addition, the MEA with GDS310 reached the maximum power density up to 0.410 and 0.396 W·cm^{-2}, which can be enhanced by approximately 31% and 27% under 5 and 20% PTFE, respectively.

References

1. S. Shimpalee, J.W. Van Zee, Numerical studies on rib & channel dimension of flow-field on PEMFC performance. Int. J. Hydrogen Energy **32**(7), 842–856 (2007)
2. Y. Prykhodko, K. Fatyeyeva, L. Hespel, S. Marais, Progress in hybrid composite Nafion®-based membranes for proton exchange fuel cell application. Chem. Eng. J. **409**(1), 27329 (2021)
3. S. Shahgaldi, A. Ozden, X. Li, F. Hamdullahpur, A scaled-up proton exchange membrane fuel cell with enhanced performance and durability. Appl. Energy **268**, 114956 (2020)
4. C.-Y. Chen, K.-P. Huang, Experimental characterization of high-temperature proton exchange membrane fuel cells under CO- and methane-containing hydrogen-rich gases. J. Appl. Electrochem. **48**(8), 911–921 (2018)
5. R.E. Rosli, A.B. Sulong, W.R.W. Daud, M.A. Zulkifley, T. Husaini, M.I. Rosli, E.H. Majlan, M.A. Haque, A review of high-temperature proton exchange membrane fuel cell (HT-PEMFC) system. Int. J. Hydrogen Energy **42**(14), 9293–9314 (2017)
6. D. Aili, D. Henkensmeier, S. Martin, B. Singh, Y. Hu, J.O. Jensen, L.N. Cleemann, Q. Li, Polybenzimidazole-based high-temperature polymer electrolyte membrane fuel cells: new insights and recent progress. Electrochem. Energy Rev **3**(4), 793–845 (2020)
7. F. Barreras, A. Lozano, V. Roda, J. Barroso, J. Martín, Optimal design and operational tests of a high-temperature PEM fuel cell for a combined heat and power unit. Int. J. Hydrogen Energy **39**(10), 5388–5398 (2014)

8. Y.-X. Wang, D.-J. Xuan, Y.-B. Kim, Design and experimental implementation of time delay control for air supply in a polymer electrolyte membrane fuel cell system. Int. J. Hydrogen Energy **38**(30), 13381–13392 (2013)
9. S.J. Peighambardoust, S. Rowshanzamir, M. Amjadi, Review of the proton exchange membranes for fuel cell applications. Int. J. Hydrogen Energy **35**(17), 9349–9384 (2010)
10. Y. Chen, K. Azizi, W. Zhang, D. Aili, S. Primdahl, L.N. Cleemann, H.A. Hjuler, Q. Li, Feasibility of using thin polybenzimidazole electrolytes in high-temperature proton exchange membrane fuel cells. Int. J. Hydrogen Energy **47**(66), 28615–28625 (2022)
11. Y. Oono, T. Fukuda, A. Sounai, M. Hori, Influence of operating temperature on cell performance and endurance of high temperature proton exchange membrane fuel cells. J. Power Sources **195**(4), 1007–1014 (2010)
12. S.K. Ryu, M. Vinothkannan, A.R. Kim, D.J. Yoo, Effect of type and stoichiometry of fuels on performance of polybenzimidazole-based proton exchange membrane fuel cells operating at the temperature range of 120–160 °C. Energy **238**, 121791 (2022)
13. K. Sasiwimonrit, W.C. Chang, The influences of catalyst combination on the high temperature proton exchange membrane fuel cell. J. Phys: Conf. Ser. **1198**(3), 032014 (2019)
14. S. Martin, J.O. Jensen, Q. Li, P.L. Garcia-Ybarra, J.L. Castillo, Feasibility of ultra-low Pt loading electrodes for high temperature proton exchange membrane fuel cells based in phosphoric acid-doped membrane. Int. J. Hydrogen Energy **44**(52), 28273–28282 (2019)
15. H. Tang, K. Geng, D. Aili, Q. Ju, J. Pan, G. Chao, X. Yin, X. Guo, Q. Li, N. Li, Low Pt loading for high-performance fuel cell electrodes enabled by hydrogen-bonding microporous polymer binders. Nat. Commun. **13**(1), 7577 (2022)
16. G. Chao, H. Tang, Q. Ju, N. Li, K. Geng, Fluorinated polybenzimidazole as binders for high-temperature proton exchange membrane fuel cells. J. Power Sources **556**, 232473 (2023)
17. R. J.Kang, Y. S. Chen, Experimental study on the effect of hydrogen sulfide on high-temperature proton exchange membrane fuel cells by using electrochemical impedance spectroscopy. Catalysts. **8**(10) (2018)
18. H. Liu, Z. Hu, J. Li, L. Xu, Y. Shao, M. Ouyang, Investigation on the optimal GDL thickness design for PEMFCs considering channel/rib geometry matching and operating conditions. Energy **282**, 128780 (2023)
19. A. Saeidfar, S. Yesilyurt, Numerical investigation of the effects of catalyst layer composition and channel to rib width ratios for low platinum loaded PEMFCs. Appl. Energy **339**, 121040 (2023)

Chapter 13
Effect of the Structure of Microporous Layer on the Performance of High-Temperature Proton Exchange Membrane Fuel Cells

Asna Widiastuti, Pongsapak Treegosol, and Yong-Song Chen

Abstract High-temperature polymer electrolyte membrane fuel cells (HT-PEMFCs) operate at 120–200°C, avoiding water flooding, tolerating CO, and improving reaction kinetics. The gas diffusion layer (GDL), which consists of the microporous layer (MPL) and carbon paper/cloth, is crucial for mass transport, heat transfer, and electricity conduction. The presence of the MPL can enhance electrochemical surface area (ECSE) and prevent phosphoric acid from leaching toward the channel. Those aspects are essential to improve performance and durability. In this work, the role of MPL in enhancing HT-PEMFC performance, with a focus on MPL structure, is investigated. The MPL with various polytetrafluoroethylene (PTFE) compositions (5, 10, and 20%) and thickness (10 to 50 μm are prepared in-house employing ultrasonic spray technology. The commercial carbon paper (GDS310, CeTech, Taiwan) is sprayed with MPL followed by 1 mg Pt cm^{-2}, while the anode is commercial carbon paper (GDL 340, CeTech, Taiwan) sprayed with 0.5 mg Pt cm^{-2}. HT-PEMFC is operated at 180 °C, with stoichiometric ratios of 1.2 and 2 for hydrogen and air, respectively. The results indicate that, for lower binder composition, increasing MPL thickness can enhance HT-PEMFC performance. However, this trend does not hold for higher binder composition. The optimal MPL structure in this study is 50 μm in thickness with 5% PTFE. The cell voltage is 0.6 V at 0.324 A cm^{-2}, while maximum power density peaks at 0.3864 W cm^{-2}. When the MPL thickness is constant, the performance decreases with increasing binder composition: a decrease of 22% at 10% PTFE and 26% at 20% PTFE. The EIS test confirms that the ohmic resistance is 2.7% and 9.7% higher at 5% PTFE compared to 10% and 20% PTFE, respectively, at a thickness of 50 μm. However, further observation is needed for 10% and 20% PTFE cases to obtain comprehensive results.

A. Widiastuti (✉) · P. Treegosol · Y.-S. Chen
Department of Mechanical Engineering, National Chung-Cheng University, Minxiong Township, Chiayi County 62102, Taiwan
e-mail: g11420146@ccu.edu.tw

C. Areeprasert et al. (eds.), *Proceedings of the International Conference on Sustainable Energy and Environmental Technology for Circular Economy*, Springer Proceedings in Energy, https://doi.org/10.1007/978-981-96-2306-8_13

Keywords High-temperature polymer electrolyte membrane fuel cell · Microporous layer · Gas diffusion layer

13.1 Introduction

High-Temperature Polymer Electrolyte Membrane Fuel Cells (HT-PEMFCs) operate at elevated temperatures between 120–200 °C [4–6], offering significant advantages over their low-temperature counterparts. These advantages include mitigating water flooding, enhancing carbon monoxide (CO) tolerance, and improving electrochemical reaction kinetics [1]. The high operational temperature prevents the condensation of water within the cell, thus eliminating the issue of water flooding, which can impede gas transport and reduce performance [1]. Furthermore, higher temperatures reduce CO poisoning of the platinum catalyst, a common issue in hydrogen fuel produced from reforming processes, by enhancing CO desorption [1, 2]. These improvements contribute to the overall efficiency and reliability of HT-PEMFCs.

A critical component in HT-PEMFCs is the gas diffusion layer (GDL), which consists of a microporous layer (MPL) and a carbon paper or cloth substrate [2, 3]. The GDL is vital in facilitating mass transport, heat transfer, and electrical conduction within the cell [4]. The MPL, in particular, is essential for enhancing the electrochemical surface area (ECSA) and preventing phosphoric acid from leaching into the flow channels [4, 5]. By optimizing the structure and composition of the MPL, the performance and durability of HT-PEMFCs can be significantly enhanced [4, 5]. The interaction between the MPL and the catalyst layer directly impacts the efficiency of reactant gas distribution and water management, which are crucial for maintaining high performance [6].

This study investigates the role of the MPL in improving HT-PEMFC performance, mainly through variations in MPL structure. Using ultrasonic spray technology, MPLs with different polytetrafluoroethylene (PTFE) compositions (5%, 10%, and 20%) and thicknesses (10 to 50 μm) were prepared and applied to commercial carbon paper substrates.

13.2 Experimental

13.2.1 Preparation of Gas Diffusion Electrode

MPL inks were prepared by dispersing carbon black (Vulcan XC-72), PTFE (60%, Aldrich, USA) in deionized (DI) water and Isopropyl Alcohol (IPA) using an ultrasonic bath and planetary mixer. The details of these cases are shown in Table 13.1. Catalyst inks in this study consist of commercial catalyst (46.7% Pt/C), Polyvinylidene Fluoride (PVDF) binder, and DMAc as solvent. All contents were mixed using

Table 13.1 The specification of the GDL in this study

5%PTFE	10% PTFE	20% PTFE
10 μm	50 μm	50 μm
30 μm		
40 μm		
50 m		

an ultrasonic bath and planetary mixer to ensure that the particle size of the mixture was sufficiently small. All of the Gas Diffusion Electrodes (GDEs) employed in this investigation had platinum loading of 0.5 mg cm^{-2} and 1 mg cm^{-2} for the anode and cathode side, respectively.

13.2.2 Preparation of MEA and Fuel Cell Test

Membrane Electrode Assembly (MEAs) were constructed in a single stack fuel cell by sandwiching the Phosphoric Acid (PA)-doped Polybenzimidazole (PBI) membrane between two Gas Diffusion Electrodes (GDEs) with an active area of 5 × 5 cm^2. The endplate, current collector, gasket, bipolar plate, GDLs, and membrane comprise a single cell. In the galvanostatic mode, polarization curves were measured while an external load applied the necessary current. The single cell was operated for 24 h at 180 °C with a constant voltage of 0.6 before I–V measurements. Tests were conducted on Hephas Energy's HS-150 fuel cell test station.

13.3 Results and Discussion

13.3.1 The Effect of Various MPL Thickness

Figure 13.1a illustrates the performance of cells with varying microporous layer (MPL) thicknesses, precisely 10, 30, 40, and 50 μm. The current densities are 0.292, 0.324, 0.336, and 0.348 A cm^{-2} for the respective MPL thicknesses at 0.6 voltage. Notably, the 50 μm MPL exhibits the highest current density of 0.348 A cm^{-2}, coupled with a maximum power density of 0.5 W cm^{-2}. This superior performance can be primarily attributed to the lowest activation loss area observed for the 50 μm MPL. This reduction in activation losses indicates more efficient catalytic activity and improved reaction kinetics, enhancing the cell's overall performance.

The results above correlate with the SEM test results in Fig. 13.2. The even coverage of the carbon paper likely increases the ECSA by maximizing the utilization of the available surface area for electrochemical reactions. This, in turn, improves the

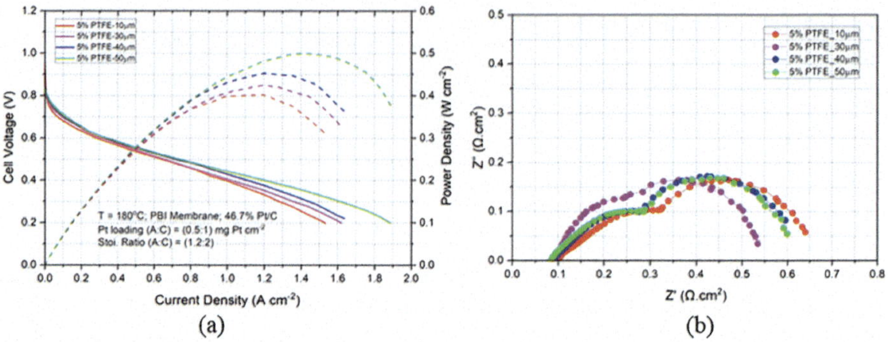

Fig. 13.1 **a** Performance curve and **b** Nyquist plots of 5%PTFE under various microporous layer thickness

cell's overall performance by increasing the number of active sites, enhancing catalyst utilization, improving reaction kinetics, and ensuring better mass transport and reduced local resistance. These factors collectively contribute to the higher current density and power density observed in the cell.

Figure 13.1b presents the Nyquist plots derived from Electrochemical Impedance Spectroscopy (EIS) measurements, providing insights into the cells' ohmic and charge transfer resistances. The ohmic resistances are 0.10202 $\Omega.cm^2$, 0.095 $\Omega.cm^2$, 0.0901 $\Omega.cm^2$, and 0.0846 $\Omega.cm^2$ for the MPL thicknesses of 10 μm, 30 μm, 40 μm, and 50 μm, respectively. Similarly, the charge transfer resistances are 0.538 $\Omega.cm^2$, 0.440 $\Omega.cm^2$, 0.506 $\Omega.cm^2$, and 0.515 $\Omega.cm^2$ for the respective MPL thicknesses. While the 50 μm MPL demonstrates the lowest ohmic resistance, signifying minimal

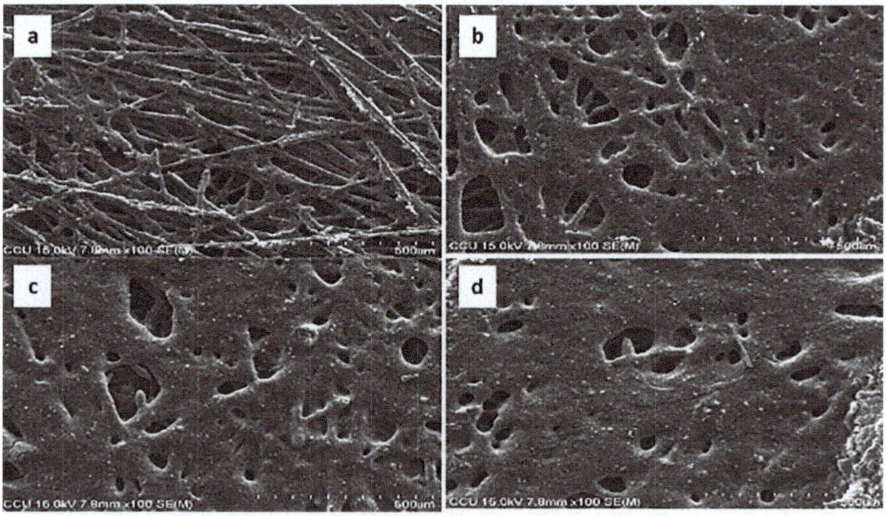

Fig. 13.2 SEM images of microporous layer surface of **a** 10 μm, **b** 20 μm, **c** 40 μm, and 50 μm

losses due to ionic transport, it exhibits a relatively high charge transfer resistance. This paradox may be due to a complex interplay between the thickness of the MPL and the resultant porosity and catalyst utilization. While a thicker MPL reduces ohmic losses by enhancing ionic conductivity, it may also lead to a suboptimal pore structure that impedes efficient charge transfer, thereby increasing the charge transfer resistance.

13.3.2 The Effect of Various PTFE Binder Percentages

Figure 13.3a presents the performance curves of cells under various percentages of PTFE binder in the microporous layer (MPL), all at a constant thickness of 50 μm. The current densities at 0.6 V are 0.348 A cm^{-2}, 0.292 A cm^{-2}, and 0.272 A cm^{-2} for MPLs with 5%, 10%, and 20% PTFE, respectively. Among these, the 5% PTFE case exhibits the highest performance, achieving a current density of 0.348 A cm^{-2} and a maximum power density of 0.5 W cm^{-2}. The superior performance of the 5% PTFE MPL can be attributed to its lowest activation loss area. Activation loss, associated with the energy required to initiate electrochemical reactions, is minimized in the 5% PTFE MPL, indicating more efficient catalytic activity and better kinetics of the electrochemical processes. This reduction in activation loss significantly contributes to the overall enhanced performance of the cell.

Figure 13.3b shows the Nyquist plots from the Electrochemical Impedance Spectroscopy (EIS) tests, providing insights into the cells' ohmic and charge transfer resistances. The ohmic resistances are 0.0846 Ω.cm^2, 0.0925 Ω.cm^2, and 0.0988 Ω.cm^2 for the 5%, 10%, and 20% PTFE MPLs, respectively. Similarly, the charge transfer resistances are 0.515 Ω.cm^2, 0.526 Ω.cm^2, and 0.563 Ω.cm^2 for the exact PTFE percentages. The 5% PTFE MPL demonstrates the lowest ohmic and charge transfer resistance values. The lower resistance in the 5% PTFE MPL can be attributed to an optimal balance between electronic conductivity and structural integrity. A lower.

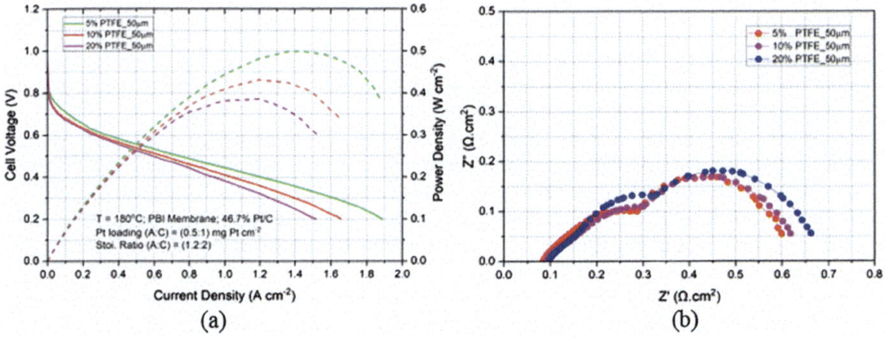

Fig. 13.3 **a** Performance curve and **b** Nyquist plots of 5%PTFE under various binder percentage

PTFE content ensures sufficient porosity for effective ion transport and minimizes the insulation effect that higher PTFE content can cause. This balance facilitates efficient charge transfer and reduces overall resistive losses, enhancing cell performance.

The percentage of PTFE binder in the MPL significantly impacts the cell's performance, even when the MPL thickness is constant. The results indicate that increasing the PTFE binder percentage from 5 to 20% leads to decreased performance, as evidenced by reduced current and power density. Increases in both ohmic and charge transfer resistances accompany this decline in performance. Higher PTFE content increases the insulation within the MPL, which hampers ionic conductivity and obstructs effective charge transfer at the electrode–electrolyte interface. Consequently, higher PTFE percentages result in more significant resistive losses, thereby diminishing the overall efficiency and performance of the cell.

13.4 Conclusions

The best performance for 5% PTFE is achieved with a 50 μm MPL thickness, showing a current density of 0.348 A cm^{-2} and a maximum power density of 0.5 W cm^{-2}. Despite having a relatively high charge transfer resistance, it has the lowest ohmic resistance among the tested samples, with values of 0.0846 $\Omega \cdot cm^2$ for ohmic resistance and 0.515 $\Omega \cdot cm^2$ for charge transfer resistance. When the MPL thickness is constant, performance decreases with increasing binder composition: a decrease of 19.8% at 10% PTFE and 27% at 20% PTFE. The EIS test confirms that at a thickness of 50 μm, the ohmic resistance is 9.3% and 16.7% higher at 5% PTFE compared to 10% and 20% PTFE, respectively. The charge transfer resistance is also 2.1% and 9.2% higher at 5% PTFE compared to 10% and 20% PTFE, respectively.

References

1. A. Zucconi, J. Hack, R. Stocker, T.A.M. Suter, A.J.E. Rettie, D.J.L. Brett, Challenges and opportunities for characterization of high-temperature polymer electrolyte membrane fuel cells: a review. J. Mater. Chem. A **12**, 8014–8064 (2024)
2. R.E. Rosli, A.B. Sulong, W.R.W. Daud, M.A. Zulkifley, T. Husaini, M.I. Rosli, E.H. Majlan, M.A. Haque, A review of high-temperature proton exchange membrane fuel cell (HT-PEMFC) system. Int. J. Hydrogen Energy **42**(14), 9293–9314 (2017)
3. H. Chun, D.-H. Kim, H.-S. Jung, J. Sim, C. Pak, Effects of gas-diffusion layer properties on the performance of the cathode for high-temperature polymer electrolyte membrane fuel cell. Int. J. Energy **48**(71), 27790–27804 (2023)
4. H. Su, C. Sita, S. Pasupathi, The effect of gas diffusion layer PTFE content on high-temperature proton exchange membrane fuel cell performance. Int. J. Electrochem. Sci. **11**(4), 2919–2926 (2016)
5. N. Bevilacqua, M.G. George, S. Galbiati, A. Bazylak, R. Zeis, Phosphoric acid invasion in high-temperature PEM fuel cell gas diffusion layers. Electrochim. Acta **257**, 89–98 (2017)

6. S. Chevalier, M. Fazeli, F. Mack, S. Galbiati, I. Manke, A. Bazylak, R. Zeis, Role of the microporous layer in the redistribution of phosphoric acid in high-temperature PEM fuel cell gas diffusion electrodes. Electrochim. Acta **212**, 187–194 (2016)
7. S. Salimi Nanadegani, E. Nemati Lay, B. Sunden, Computational analysis of the impact of a microporous layer (MPL) on the characteristics of a high-temperature PEMFC. Electrochim. Acta **333**, 135552 (2020)

Chapter 14
Impact of Building Shading on Solar Energy Potential of Rural Clusters: A Case Study of Nanjing, China

Jiang Liu and Changhai Peng

Abstract Rural building clusters have significant prospects for solar energy application. However, the unclear impact of building shading on solar radiation distribution impedes their effective utilization. To address this issue, we conducted case studies in three villages with diverse morphologies in Nanjing. Employing a 12.5 m resolution digital elevation model (DEM), 3D building models, and hourly radiation simulation, we investigated the annual accumulated solar irradiance of these villages with and without building shading. The results showed that the impact of mutual shading is more substantial on facades than on rooftops. Without the shading, the façade annual accumulated solar radiation is 29.8–55.5% higher. While for rooftops, this ratio is only 2.9–4.7%. For different oriented surfaces, west-facing rooftops and facades are the most sensitive. Neglecting building shading leads to overestimations of 4.0–20.8% and 37.0–84.8% for the solar irradiance on them, respectively. Furthermore, the morphological disparity of villages would interfere with the impact of building shading. This study contributes to a more accurate assessment of solar energy potential in rural areas, which is vital for constructing a sustainable society.

Keywords Rural clusters · Building shading · Solar energy potential · China

14.1 Introduction

Utilizing solar energy is regarded to be a critical solution for curbing climate change, due to the abundance, accessibility, and safety of this energy. Particularly in rural areas that have a large number of low-rise buildings and ample vacant land, solar energy

J. Liu (✉) · C. Peng
School of Architecture, Southeast University, Nanjing 210096, China
e-mail: liujiang@seu.edu.cn

C. Peng
Key Laboratory of Urban and Architectural Heritage Conservation (Southeast University), Ministry of Education, Beijing, China

C. Areeprasert et al. (eds.), *Proceedings of the International Conference on Sustainable Energy and Environmental Technology for Circular Economy*, Springer Proceedings in Energy, https://doi.org/10.1007/978-981-96-2306-8_14

applications possess promising prospects. In recent years, numerous studies have paid attention to this field, especially for building-integrated solar energy systems [1, 2]. For instance, Sun et al. [3] utilized high-resolution satellite images to identify the flat and pitched rooftops of villages in northern China and estimated the corresponding PV power potential. Zhong et al. [4] and Yang et al. [5] used satellite images and vector-building footprints to reveal the possible rural rooftop PV power capacity of coastal cities in southeastern China. However, it is necessary to point out that due to the lack of 3D building models, the impact of building shading is generally ignored in these studies. Although the low-rise characteristic of rural buildings can support this simplification to a certain extent, its precise consequences have not been unveiled. In light of this, this study employed three villages in Nanjing as cases and preliminarily quantified the impact of building shading on the solar energy potential of rooftops and facades with different orientations.

14.2 Study Area and Methodology

14.2.1 Study Area

This study focused on Nanjing, a metropolis in eastern China. Due to the subtropical monsoon climate, Nanjing has an annual average temperature and precipitation of 15.4 °C and 1200 mm, respectively, with the sunlight hours being about 2000 h per year. By the end of 2021, Nanjing accommodates a residential population of 942.3 million and spans an area of 6587.04 km^2 [6]. This huge population incurred a great demand for energy. For instance, in 2021, Nanjing's electricity usage reached 68.4 TWh, accounting for about 9.6% of the total electricity use of Jiangsu province (the higher-level administrative region encompassing Nanjing) [7]. This exerts great stress on the city's sustainable development and drives it to seek a clean energy transition.

Among various clean energy, solar energy was deemed to be the promising one for this city, especially in its rural areas. With a nearly 1.96 million household population and 66.5 m^2 per capita floor area in these regions, rural buildings of Nanjing have approximately 130 km^2 of floor area, which provides substantial space for solar energy technologies (e.g., PV panels) application. For example, according to the report of [5], rural buildings in this city can generate a total of 6.83 TWh annual rooftop PV power, taking up 10.8% of the city's electricity usage in 2021. Therefore, selecting Nanjing as a study area is valuable for exploring rural solar resources to achieve sustainable development.

In this work, three villages distributed in different geographical locations were selected as case studies, as shown in Fig. 14.1a, b. With the 12.5 m resolution digital elevation model (DEM) downloaded from Alaska Satellite Facility (ASF, https://asf.alaska.edu/) and the three-dimensional (3D) building models developed according to [8], their morphological characteristics were described in Fig. 14.1c. It can be

ID	Name	TFA /m²	MBO	ATV /m	Remark
I	Zhoulang village	17296.44	21.7° south by east	0.73	Min terrain variation
II	Beisheng village	32874.59	6.32° south by east	3.00	Moderate terrain variation
III	Danian village	11874.68	9.48° south by east	5.59	Max terrain variation

Note: TFA: total floor area, MBO: Main building orientation, ATV: average terrain variation

Fig. 14.1 The selected three villages. **a** Location, **b** Satellite image, **c** 3D characteristic

observed that the total floor area of these villages ranges from 11,874.68 to 32,874.59 m^2, while the main building orientations are all southeast. The average terrain variation within a village varies from 0.73 to 5.59 m. These discrepancies allow us to rigorously analyze the impacts of building shading on solar energy potential, particularly for considering the interference of morphology.

14.2.2 Solar Radiation Simulation with and Without Building Shading

With the 12.5 m resolution DEM and 3D building models, this study employed Ladybug Tools to calculate the annual accumulated solar radiation distribution of villages. Ladybug Tools is a specialized integrated platform for building physics analysis. It includes four main modules: Ladybug, Honeybee, Dragonfly, and Butterfly. Among these, Honeybee is dedicated to simulating building energy performance, e.g., air conditioning system design and indoor daylight assessment, by invoking related simulation engines, such as EnergyPlus and Radiance. The backward ray tracing method adopted in Radiance enables Honeybee can accurately calculate the solar radiation distribution on building exterior surfaces in high spatiotemporal resolution. Here, we utilized Honeybee to conduct the simulation.

Initially, the DEM and 3D building models were fed into Rhino, a 3D design software. Then the rooftops and facades of these building models were discretized into parallel grids. After importing the typical weather file of Nanjing, Radiance was invoked to simulate the distribution of solar radiation on each grid with a timestep of 1 h. By aggregating the results of these grids, we ultimately derive the annual accumulated solar radiation distribution of rooftops and facades. Notably, this process was repeated twice, one for considering building shading by taking each building as the context, while the other was for ignoring building shading where only topography

Table 14.1 Parameter settings of solar radiation simulation

Parameters	Settings	Descriptions
Grid size	2 × 2 m	The sizes of discretized grids
Reflectance	Rooftops: 0.48, Facades: 0.35, Ground: 0.26	The proportion of solar radiation reflection
Ambient bounce	3	The maximum number of sky diffuse bounces within a village
Recipe type	Rfluxmtx	The recipe adopted to calculate solar radiation
Detail level	High	The quality of model parameters

was set to the context. Detailed settings of these simulations are summarized in Table 14.1.

14.3 Results and Discussions

14.3.1 Annual Accumulated Solar Radiation Distribution in Rural Building Clusters

Figure 14.2 shows the annual accumulated solar radiation on rooftops and facades of each village when considering building shading. It can be seen that the total amounts of radiation on rooftops and facades are close to each other with a ratio of approximately 50% versus 50%. In village I, the annual accumulated solar irradiances of rooftops and facades are 13.86 and 12.22 GWh (1 GWh equals 10^6 kWh), respectively. While for village III these values are 9.63 and 8.52 GWh, respectively. This depicts that rooftops and facades of rural building clusters have nearly equivalent prospects for solar energy application. Furthermore, among differently oriented building surfaces, south- and north-facing rooftops and south- and west-facing facades have the highest annual accumulated solar irradiance, they jointly contribute to over 60% of the total solar radiation in each village, indicating they should be given priority when exploring rural solar resources.

14.3.2 Impacts of Building Shading on Rooftop and Façade Solar Energy Potential

Figure 14.3 illustrates the annual accumulated solar radiation distribution of each village with and without building shading. It can be seen that the impact of building shading on facades is higher than on rooftops. In Fig. 14.3a, after excluding building shading, the annual accumulated solar radiation of total facades increased by 2.54 to

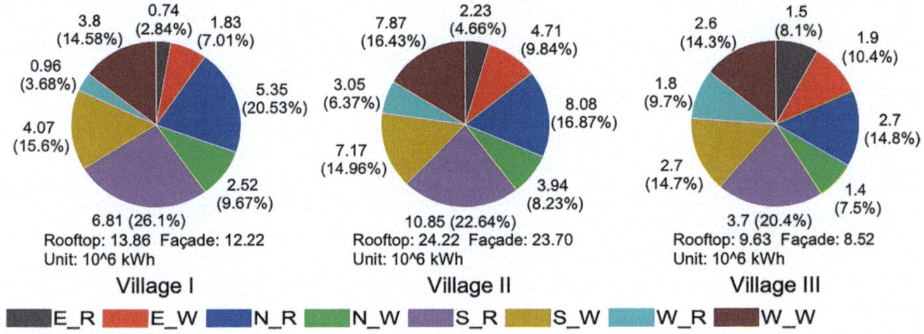

Fig. 14.2 Annual accumulated solar radiation of three villages. (R: rooftop, W: façade, E: east-facing, N: north-facing, S: south-facing, W: west-facing)

8.31 GWh, whilst the corresponding increment for rooftops is only 0.31 to 0.71 GWh. Figure 14.3b depicts the relative variations of solar radiation of them. In line with Figure 14.3a, the growth degree of facades is more significant. When building shading is ignored, solar energy potential on this type of surface rises by 29.8% to 55.5%, far surpassing the ratio of 2.9% to 4.7% of rooftops. Additionally, it is observed that village type has an interference with the impact of building shading. In Village I, the overestimated percentage for façade solar irradiance exceeds 50%, while for Village II and III, the overestimations are less than 40%. This may be attributed to their morphological disparities. As shown in Fig. 14.1c, compared to Village II and III, Village I holds more easterly orientation and flat terrain. Therefore, in future evaluation of the effect of building shading, the impact of village morphology should be taken into account.

Specific to orientations of rooftops and facades, Fig. 14.3c, d show that west-facing surfaces are most easily influenced. In the three villages, when building shading is ignored, the annual accumulated solar radiation of west-facing rooftops and facades is overestimated by 4.0% to 20.8% and 37.0% to 84.8%, respectively. Particularly for Village I, the overestimation error is highest. Hence, when utilizing west-facing surfaces to deploy solar energy systems, the impact of building shading must be considered, even if they are rooftops.

Finally, this study has some limitations that need to be addressed. First, vegetation in rural areas (e.g., trees and grass) was not included in these 3D village models, causing a certain error in their impact evaluation. Second, a higher temporal-resolution assessment (e.g., monthly and daily) on building shading impact should be implemented to guide the deployment of solar energy systems in practice.

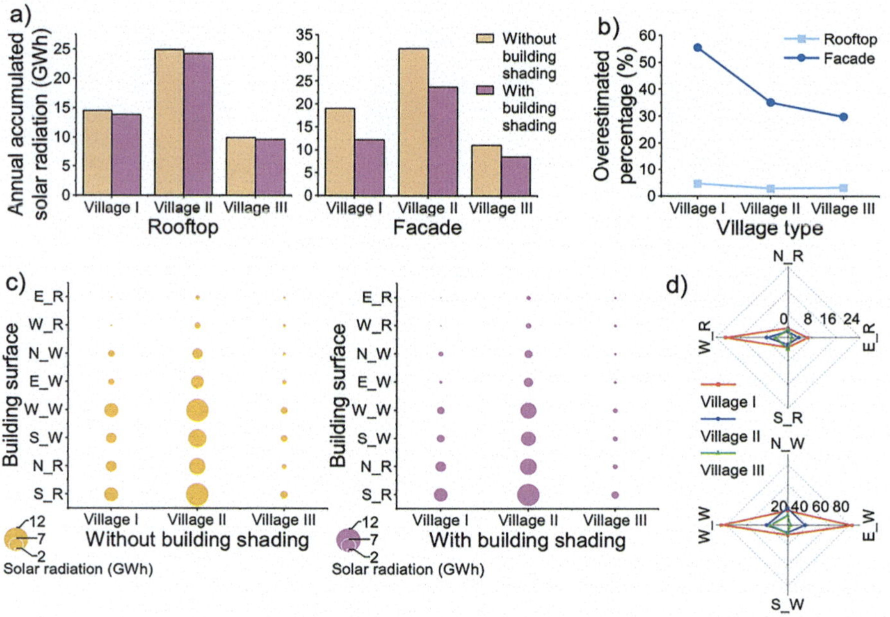

Fig. 14.3 Annual accumulated solar radiation of three villages with and without building shading. **a** Absolute variations of rooftops and facades, **b** Relative variations of rooftops and facades, **c** Absolute variations for different orientations, **d** Relative variations for these orientations

14.4 Conclusions

With the studies of three villages in Nanjing, this work found that the impact of building shading on solar energy potential is more significant on facades than on rooftops. Ignoring this shading, the annual accumulated solar radiation of facades would be overestimated by 29.8% to 55.5%, while for rooftops this error is only 2.9% to 4.7%. Among various oriented surfaces, west-facing rooftops and facades are the most sensitive. The overestimations for them can reach maximums of 20.8% and 84.8%, respectively. Additionally, village morphology exerts interferences with building shading. Therefore, their effect should be considered in future studies.

Acknowledgement This work was supported by the National Natural Science Foundation of China (Grant NO. 52378011).

References

1. A. Zhao, Y. Jiao, W. Quan, Y. Chen, Net zero carbon rural integrated energy system design optimization based on the energy demand in temporal and spatial dimensions. Renew. Energy **222**, 119818 (2024)

2. S.S. Hosseini Dehshiri, B. Firoozabadi, Building integrated photovoltaic with hydrogen storage as a sustainable solution in Iranian rural healthcare centers. Energy Conv. Manag. **314**, 118710 (2024)
3. T. Sun, M. Shan, X. Rong, X. Yang, Estimating the spatial distribution of solar photovoltaic power generation potential on different types of rural rooftops using a deep learning network applied to satellite images. Appl. Energy **315**, 119025 (2022)
4. T. Zhong et al., A city-scale estimation of rooftop solar photovoltaic potential based on deep learning. Appl. Energy **298**, 117132 (2021)
5. Y. Yang et al., Whether rural rooftop photovoltaics can effectively fight the power consumption conflicts at the regional scale—a case study of Jiangsu Province. Energy Build. **306**, 113921 (2024)
6. Nanjing Bureau of Statistics, *Nanjing Statistical Yearbook 2022* (China Statistics Press, Beijing, 2022)
7. Jiangsu Bureau of Statistics, *Jiangsu Statistical Yearbook 2022* (China Statistics Press, Beijing, 2022)
8. J. Liu et al., A novel approach for assessing rooftop-and-facade solar photovoltaic potential in rural areas using three-dimensional (3D) building models constructed with GIS. Energy **282**, 128920 (2023)

Part IV
Analysis of Energy and Environmental Systems

Chapter 15
Secure Data Analytics in Smart Grids: Preserving Privacy and Enabling Advanced Monitoring

Mayur Rele, Anitha Julian, Dipti Patil, Gerardine Immaculate Mary, R. Ramyadevi, and M. Udaya Krishnan

Abstract Integrating developing technology and data analytics in smart grids has raised apprehensions over security and privacy. This review article examines the techniques and procedures employed in secure data analytics for smart grids. It uses these approaches to improve grid monitoring, discover anomalies, and make accurate forecasts. It also emphasizes the need to protect data privacy and maintain confidentiality throughout the process. The study encompasses several focal areas: homomorphic encryption, secure multi-party computation, and specific data analytics about smart grids. This study examines the safeguarding of encrypted grid data using machine learning techniques. It explores the many obstacles encountered in the design and execution of such measures within the context of the smart grid domain. Through the synthesis and critical examination of pertinent scholarly literature, this research elucidates the capacity of secure data analytics to detect anomalies, forecast grid patterns, and enhance decision-making procedures. Additionally, the analysis underscores the presence of undiscovered domains and potential avenues for future research in safe data analytics for smart grids. This extensive review is a significant reference for researchers, professionals, and politicians seeking to understand and enhance the methods that guarantee the privacy and security of smart grid operations.

M. Rele (✉)
IT and Cybersecurity, Parachute Health, Princeton, NJ, USA
e-mail: mayur.rele@parachutehealth.com

A. Julian · R. Ramyadevi · M. Udaya Krishnan
Saveetha Engineering College, Thandalam, Chennai, India

D. Patil
Graduate School of Business, University of Cumberlands, Williamsburg, KY, USA

G. I. Mary
School of Electronics Engineering, Vellore Institute of Technology Vellore, Vellore, Tamil Nadu, India

C. Areeprasert et al. (eds.), *Proceedings of the International Conference on Sustainable Energy and Environmental Technology for Circular Economy*, Springer Proceedings in Energy, https://doi.org/10.1007/978-981-96-2306-8_15

Keywords Secure data analytics · Smart grids · Grid monitoring · Anomaly detection · Prediction · Data privacy · Data confidentiality · Cryptographic techniques · Homomorphic encryption

15.1 Introduction

The energy industry has undertaken a transition due to the increasing adoption of cutting-edge technology and data-driven analytics in smart grids, which have transformed grid management, energy distribution, and the absorption of renewable energy sources [1]. These developments have resulted in various advantages, such as enhanced sustainability, increased efficiency, and optimized grid operations. Nonetheless, this digital revolution carries several crucial security and privacy vulnerabilities that must be addressed [2]. The interconnected components of smart grids include smart meters, sensors, and communication networks. These components capture real-time information regarding energy consumption, grid operation, and environmental conditions [3]. This abundance of data provides opportunities for capacity forecasting, demand response programs, and the incorporation of renewable energy [4]. Due to the sensitivity of the collected data and the infrastructural flaws of the smart grid, hostile attacks, privacy violations, and unauthorized access are, however, possible [5]. Traditional security solutions, such as firewalls and access restrictions, must be revised to address the sophisticated security challenges smart grids present [6]. Due to the dynamic nature of the grid, the enormous volume of data generated, and the need for swift decision-making, novel approaches for secure data analytics are required [7]. Ensuring the privacy and confidentiality of grid data, cryptographic techniques, and machine learning algorithms offer intriguing solutions for enhanced grid monitoring, anomaly detection, and prediction [8]. Two cryptographic techniques, secure multi-party computation, and homomorphic encryption, provide mechanisms for performing calculations on encrypted data while maintaining the confidentiality of private data even during analysis [9]. Data processing and analysis can be conducted securely and anonymously using these methods. Combining these cryptographic approaches with machine learning algorithms [10] makes it possible to decrypt encrypted grid data while safeguarding the privacy of individual users and preserving the confidentiality of sensitive grid data. The imperative need to address security and privacy issues in the context of smart grid data analytics prompted this review study. Researchers, practitioners, and regulators can develop and implement safe and privacy-preserving solutions for smart grid operations if they thoroughly comprehend the current methodology and advancements in secure data analytics. These initiatives have the potential to strengthen cybersecurity defenses, enhance grid resilience, and cultivate greater trust among energy industry stakeholders. Several research projects have conducted extensive research on this topic to reduce cyberattack risks and preserve the security and privacy of smart grid systems.

These studies have focused on various topics, such as developing secure communication protocols, reliable authentication methods, and efficient encryption techniques. This review article seeks to consolidate available knowledge by analyzing and synthesizing diverse research results to shed light on cutting-edge approaches and advancements in safe data analytics for smart grids. This study investigates in-depth secure data analytics methodologies for smart grids and the interaction between machine learning and cryptography.

15.2 Security and Privacy Challenges in Smart Grid

A smart grid system can increase the efficiency of power generation, transmission, and distribution. Utilizing various power-producing sources, including renewable energy, enables end users to have an effective and dependable electricity supply [11]. Without advanced metering infrastructure (AMI), which consists of smart meters for real-time energy monitoring and support for demand response programs, there can be no smart grid [12]. Strong communication networks provide smooth information exchange between grid components, including real-time data transfer and monitoring [13]. Energy storage devices are essential for accommodating renewable energy sources and maintaining system dependability. By modifying parameters, analyzing data, and forecasting demand, intelligent control systems and data analytics techniques can be used to optimize grid operations [14]. One of the many advantages of incorporating these elements and technologies into a smart grid system is the increased grid reliability, decreased energy losses, increased integration of renewable energy, increased resilience against outages, support for electric vehicles, and empowered consumer participation in energy management [15]. In addition to promoting energy efficiency and lowering greenhouse gas emissions, smart grid technologies also support the global transition to a more robust and sustainable energy future. Only two of the many advantages of modernizing electrical distribution networks to what are known as "smart grids" and incorporating cutting-edge technologies are increased efficiency and the inclusion of renewable energy sources. However, they also bring up several issues related to privacy and security. Cybersecurity is a severe concern because smart grids depend on information and communication technology (ICT) systems that hackers might attack [16]. Due to the enormous volumes of data that smart grids collect and the need to safeguard them from unwanted access and misuse, data privacy is another crucial concern [17]. The grid's spread structure makes it simpler for attackers to access additional system points. The risks to data security are further exacerbated by insider threats and outdated systems' weaknesses [18]. Maintaining the security and dependability of smart grids is difficult due to the need for standardized security measures and the requirement for resilience and incident response techniques [19]. Strong encryption, routine software upgrades, extensive security reviews, thorough employee training, strategic partnerships, industry-wide standards, detailed privacy effect analysis, and privacy-enhancing technology are the most efficient remedies for these issues. The architecture of the smart grid infrastructure should incorporate security and privacy from the start, and a security-conscious culture should be promoted.

15.3 Overview of Cryptographic Methods in Smart Grid Data Analytics

Cryptographic techniques are indispensable for ensuring the confidentiality and authenticity of data analytics in smart grids [20]. Smart grid systems collect, analyze, and manage massive quantities of data from various sources, including sensors, meters, and other intelligent devices [21]. Personal information (PII), energy consumption patterns, and user behavior are a few examples of sensitive data that fall under this category. Cryptography safeguards the confidentiality, honesty, and integrity of this information. Encryption is a vital component of the cryptography procedure [22] when analyzing data from a smart grid. Encryption is the process of rendering data unintelligible using a cryptographic key and an encryption algorithm as shown in Fig. 15.1. The capacity to access the data is limited to individuals who possess the appropriate decryption key. Encryption is commonly employed as a security measure to safeguard against unauthorized access or interception in data transmission and storage. Furthermore, homomorphic encryption and analogous techniques enable computations to be executed on encrypted data sans decryption, thereby safeguarding users' anonymity while scrutinizing the data. A variety of cryptographic methods have been suggested as a means of safeguarding smart grid data analytics; each has its own set of benefits and drawbacks. Homomorphic encryption allows for the computation of encrypted data without the need for decryption, thereby offering privacy protection. Conversely, the substantial computational expense that real-time processing necessitates is not always advantageous. By contrast, Secure Multi-Party Computation (MPC) facilitates anonymous collaboration among remote parties during the computation of calculations. Despite the potential cost of communication, this renders it a more scalable option. For instance, homomorphic encryption provides the highest level of privacy assurances, but it is computationally costly. Secure enclaves and TEEs, on the other hand, offer speedier processing but may be susceptible to physical attacks or advanced malware that targets the hardware.

Digital signatures [23] are another cryptographic technique employed in smart grid data analytics. The verification of data authenticity and integrity can be facilitated through the utilization of digital signatures. The generation of a signature that is affixed to data is accomplished through the utilization of a private key. Utilizing the appropriate public key makes it possible to verify the signature and ascertain that the information has not undergone any modifications while being transmitted or stored. The utilization of digital signatures is imperative in establishing trust and preventing unauthorized data modifications, ultimately guaranteeing the authenticity and integrity of smart grid analytics. Cryptographic protocols, including secure key exchange and communication channels, are essential in data analytics for smart grids

Fig. 15.1 Architecture of system encryption

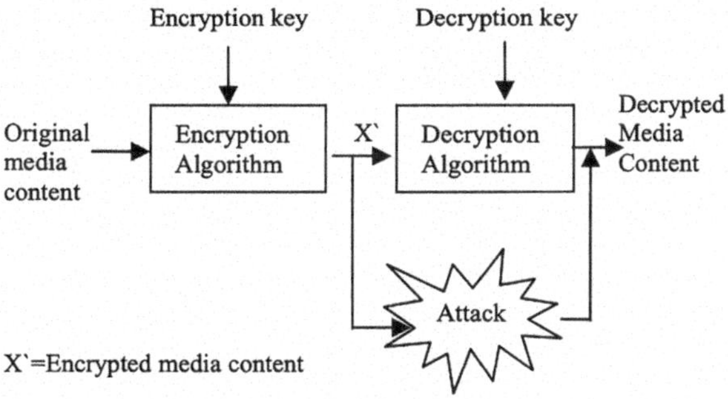

[24]. The protocols above establish secure channels to facilitate data transmission, ensuring the confidentiality and integrity of information by preventing unauthorized access and tampering. The Transport Layer Security (TLS) protocol is commonly utilized to ensure secure communication between various components of the smart grid, such as meters and data aggregators. To establish secure connections, various techniques such as encryption, digital certificates, and authentication are employed. The use of cryptographic methods in data analytics for smart grids has been extensively explored in various scholarly publications. Ghosal's study [25] explores using cryptographic algorithms to facilitate secure and efficient data transmission within the smart grid. The investigation pertains to using encryption techniques to safeguard the confidentiality of information during its transmission. A recent study [25] highlights the significance of cryptographic techniques in upholding confidentiality during the consolidation of confidential information and safeguarding data aggregation in the smart grid context. The authors of reference [26] have proposed a secure system for aggregating and computing data in smart grid communication. The scheme above employs cryptographic techniques to ensure the confidentiality and integrity of data during the aggregation process. The authors in [27] have proposed an approach for privacy-preserving data aggregation in the smart grid, which utilizes homomorphic encryption. This approach enables computations to be performed on encrypted data without requiring decryption. Presents a novel approach to safeguarding privacy during data aggregation in the smart grid. This approach involves the integration of blockchain technology and homomorphic encryption. The proposed strategy ensures secure and transparent data aggregation while safeguarding the privacy of individual data providers. Wang, Jiang, and Shen have proposed a data aggregation technique that employs cryptographic mechanisms to encrypt data during transmission. This approach ensures the confidentiality and integrity of aggregated data in smart grid connections. The aforementioned scholarly works tackle confidentiality, safeguarding measures, and efficient data consolidation. In unison, they propel the domain of cryptographic methodologies into intelligent power grid data analysis.

15.4 Integration of Cryptographic Methods with Machine Learning Algorithms

The convergence of cryptographic techniques and machine learning algorithms presents an intriguing confluence of two powerful technologies that can potentially augment data-centric systems' security, confidentiality, and dependability. Cryptographic techniques offer effective means of safeguarding confidential information, while machine learning algorithms enable sophisticated analysis and informed decision-making. The integration of these two domains has given rise to innovative methodologies for addressing the predicaments of safeguarding data confidentiality while capitalizing on the advantages of machine learning. Privacy-preserving machine learning is a crucial research domain that utilizes cryptographic methods to safeguard confidential information during the training and inference stages. Secure multi-party computation (MPC) and homomorphic encryption are viable techniques that enable collaborative machine learning model training on encrypted data while maintaining data privacy as shown in Fig. 15.2. This facilitates the ability of data owners to contribute their data to a shared model while maintaining confidentiality. The potential privacy and security benefits of integrating cryptographic techniques with machine learning are yet to be realized by smart grid systems. While homomorphic encryption does enable the training of machine learning models on encrypted data, the significant computational inefficiencies it generates render it unsuitable for real-time anomaly detection or grid monitoring. Although Secure MPC enables decentralized model training while safeguarding data privacy, its substantial communication cost may result in disruptions in large-scale grid operations. Despite its potential to protect user privacy, federated learning contains numerous security vulnerabilities. Although differential privacy strategies have been demonstrated to mitigate the likelihood of leakage during model updates, recent research has demonstrated that models are still susceptible to membership inference assaults and adversarial manipulation. Hybrid approaches that integrate secure enclaves with machine learning have been proposed as a solution to the problem of federated learning environments lacking cryptographic guarantees. This approach would ensure the security of users' models and their personal information. Additionally, each cryptographic method has its own set of advantages and disadvantages. Digital signatures and authentication techniques may introduce latency that may pose issues for systems that require real-time data analysis, despite the fact that they enhance trust. As a result, in order to maintain security and circumvent these restrictions while integrating cryptography with machine learning, more complex methods are necessary. Despite advancements, there is still a void in the provision of end-to-end security in circumstances that necessitate the rapid processing of vast quantities of data.

Furthermore, cryptographic techniques may be employed to authenticate and ensure the integrity of machine learning models. Technologies such as digital signatures and secure enclaves offer to authenticate a model's integrity throughout its entire lifespan, ensuring that it has not been tampered with or jeopardized. The measure above ensures the model's credibility and safeguards against potential adversarial

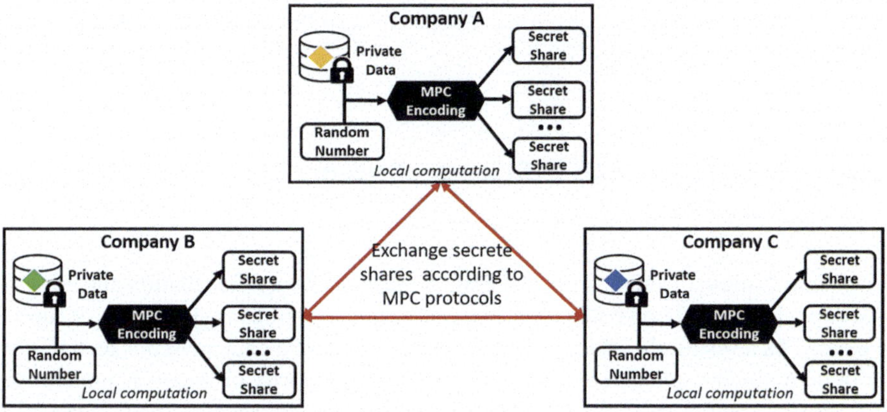

Fig. 15.2 Secure multi-party computation

attacks that may compromise the integrity of the outcomes. By utilizing secure aggregation and differential privacy, two cryptographic techniques, it becomes feasible to consolidate model updates from various origins while maintaining the confidentiality of individual data. This enables collaborative model training between businesses and individuals while maintaining data privacy. Several research articles have explored the integration of cryptographic techniques with machine learning algorithms as a means of addressing concerns related to privacy, security, and trust. The reference authors propose a technique for constructing deep learning models with differential privacy to safeguard sensitive data confidentiality during model training. Presented a valuable technique for ensuring secure aggregation in federated learning. This method enables the incorporation of model updates from multiple parties while preserving data privacy. Scholarly literature has examined membership inference attacks on machine learning models. These authors have emphasized the importance of cryptographic methods in safeguarding the anonymity of individual data contributors. The presented techniques and conducted performance evaluations for federated learning utilizing differential privacy to ensure privacy preservation in collaborative learning situations. This also presents techniques for constructing language models that ensure privacy, specifically emphasizing differential private recurrent language models. A comprehensive exposition of secure multi-party computation in machine learning. This includes an extensive examination of cryptographic protocols and their potential applications in preserving data confidentiality during collaborative computations. The studies in question enhance the fusion of cryptographic methodologies and machine learning algorithms, thereby enabling the preservation of privacy in machine learning, secure data aggregation, and protection against privacy breaches. This, in turn, enhances the reliability of data-centric systems. Integrating cryptographic techniques can enhance the accountability and transparency of machine learning systems. Users can verify machine learning algorithms' outcomes and judgments using cryptographic protocols that offer verifiable calculations and proof of accuracy. This mitigates apprehensions regarding partiality or malicious conduct and enhances trust in the results.

15.5 Effectiveness and Efficiency of Secure Data Analytics Method

Within the domain of smart grids, the optimization of energy distribution and administration relies heavily on collecting, analyzing, and applying large amounts of sensitive data. As a result, the efficacy and efficiency of secure data analytics techniques are of paramount significance. Implementing strong security protocols is imperative to ensure data analytics' confidentiality, reliability, and integrity within the smart grid system. The assessment of secure data analytics methods in the smart grid requires evaluating their ability to prevent data intrusions, uphold data confidentiality, and enable accurate analysis and decision-making. The implementation of encryption is imperative in ensuring the safeguarding of data analytics for smart grids. Encryption methods prevent unauthorized access and ensure the confidentiality of data by encoding sensitive information, such as user data or energy consumption patterns, during transmission and storage processes. Strong encryption methods, such as symmetric and asymmetric encryption algorithms, can effectively safeguard against unauthorized access and monitoring, thus improving the efficacy of secure data analysis in the smart grid. Secure computation techniques are critical in preserving privacy and enabling efficient data analysis within the smart grid infrastructure. Furthermore, these techniques are essential for protecting user privacy. Homomorphic encryption allows for computations on encrypted data without decryption. The functionality above facilitates the process of conducting data analytics while upholding the confidentiality of sensitive information, ensuring efficient and secure analysis of smart grid data. Incorporating access controls and authentication mechanisms is a crucial aspect of implementing secure data analytics methodologies in the smart grid context. Organizations can regulate access to sensitive data and prevent unauthorized manipulation or disclosure by implementing access controls based on roles, privileges, or user attributes. The efficacy of secure data analytics can be enhanced by implementing robust authentication protocols, such as public key infrastructure (PKI) or secure multi-factor authentication. These protocols restrict access to smart grid data solely to authorized personnel. Efficiency is a crucial aspect to consider in analyzing smart grid data. To ensure efficient decision-making and energy management, implementing smart grids necessitates utilizing real-time or nearly real-time analysis of extensive data sets. The achievement of secure data analytics necessitates implementing methods that effectively balance computational efficiency with privacy guarantees. The enhancement of secure data analytics in the smart grid is facilitated by deploying effective data processing architectures and refining cryptographic algorithms. Furthermore, the incorporation of sophisticated technologies such as secure enclaves, trusted execution environments (TEE), and hardware security modules (HSM) has the potential to enhance the efficacy and productivity of secure data analytics in the smart grid. The technologies above offer secure and segregated settings for handling confidential information, reducing the likelihood of unapproved entry while maintaining optimal efficiency. The selected scholarly articles investigate the efficiency and potency of secure data analytics techniques within the context of smart grids. This presents a comprehensive examination of privacy-preserving data analytics, emphasizing the challenges and techniques specific to the smart grid. The aforementioned edited volume comprehensively analyzes the security and privacy issues that arise in smart grids, encompassing data analytics. This article thoroughly examines secure data analytics within the context of smart grids, encompassing a diverse array of techniques and methodologies. Privacy-preserving data analytics in the context of smart grids is the subject of further investigation, underscoring the significance of safeguarding privacy. Presenting a data aggregation scheme that ensures security and privacy in smart grid communications while providing effectiveness and safety. Finally, it offers a comprehensive analysis of privacy-preserving data analytics in the context of smart grids. This source delves into privacy-related issues and explores diverse methodologies and technologies. The research above contributes to secure data analytics in the smart grid domain. They address important topics such as privacy protection, efficiency considerations, and comprehensive surveys, which collectively enhance our understanding of effective and efficient methods in this area.

15.6 Conclusion

In conclusion, it is crucial to prioritize incorporating safe data analytics methodologies to uphold the confidentiality, integrity, and efficiency of data-centric activities inside smart grids. The present research has conducted an in-depth examination of prominent approaches, including homomorphic encryption, safe multi-party computation, and machine learning methodologies. These methods have been elucidated for their significance in safeguarding confidential data and facilitating secure data analysis in intelligent power grids. The study emphasizes crucial elements such as the protection of encrypted grid data, the identification of anomalies, the prediction of grid behavior, and the issues associated with the design and operation of smart grids. Additionally, this study has identified certain domains in which research is deficient and identified potential avenues for future investigation. These include balancing maintaining confidentiality with maximizing the utility of data, broadening the breadth of inquiry, and establishing standardized protocols. Promising avenues for prospective study encompass the investigation of advanced cryptographic techniques, the development of efficient protocols, the initiation of standardization endeavors, and the exploration of emergent technologies. By conducting in-depth studies into these less explored domains and potential avenues for future research, the field of secure data analytics in smart grid systems has the potential to make significant progress in attaining enhanced levels of effectiveness, efficiency, and reliability. The collaborative endeavor described can potentially result in the advancement of smart grid systems that prioritize both security and privacy. This aligns with the overarching goal of improving the safety and confidentiality of smart grid operations.

References

1. T. Ahmad, R. Madonski, D. Zhang, C. Huang, A. Mujeeb, Data-driven probabilistic machine learning in sustainable smart energy/smart energy systems: key developments, challenges, and future research opportunities in the context of smart grid paradigm. Renew. Sustain. Energy Rev. **160**, 112128 (2022)
2. K. Kimani, V. Oduol, K. Langat, Cyber security challenges for IoT-based smart grid networks. Int. J. Crit. Infrastruct. Prot. **25**, 36–49 (2019)
3. H. Shahinzadeh, J. Moradi, G.B. Gharehpetian, H. Nafisi, M. Abedi, IoT architecture for smart grids, in *2019 International Conference on Protection and Automation of Power System (IPAPS)* (IEEE, 2019), pp. 22–30
4. B.N. Silva, M. Khan, K. Han, Futuristic sustainable energy management in smart environments: a review of peak load shaving and demand response strategies, challenges, and opportunities. Sustainability **12**(14), 5561 (2020)
5. M.Z. Gunduz, R. Das, Cyber-security on smart grid: threats and potential solutions. Comput. Netw. **169**, 107094 (2020)
6. C. Ruland, J. Sassmannshausen, Firewall for attribute-based access control in smart grids, in *2018 IEEE International Conference on Smart Energy Grid Engineering (SEGE)* (IEEE, 2018), pp. 336–341
7. J. Moradi, H. Shahinzadeh, H. Nafisi, M. Marzband, G.B. Gharehpetian, Attributes of big data analytics for data-driven decision making in cyber-physical power systems, in *2020 14th International Conference on Protection and Automation of Power Systems (IPAPS)* (IEEE, 2019), pp. 83–92
8. L. Cui, Y. Qu, L. Gao, G. Xie, S. Yu, Detecting false data attacks using machine learning techniques in smart grid: a survey. J. Netw. Comput. Appl. **170**, 102808 (2020)
9. D. Das, Secure cloud computing algorithm using homomorphic encryption and multi-party computation, in *2018 International Conference on Information Networking (ICOIN)* (IEEE, 2018), pp. 391–396
10. A. Ziller, A. Trask, A. Lopardo, B. Szymkow, B. Wagner, E. Bluemke, G. Kaissis, *Pysyft: A Library for Easy Federated Learning. Federated Learning Systems: Towards Next-Generation AI* (2021), pp. 111–139
11. M.M. Kumbhar, P.R. Narvekar, A.B. Nandgaonkar, S.L. Nalbalwar, Smart grid: advanced electricity distribution network. IOSR J. Eng. (IOSRJEN) **2**(6), 23–29 (2012)
12. A. Kailas, V. Cecchi, A. Mukherjee, A survey of communications and networking technologies for energy management in buildings and home automation. J. Comput. Netw. Commun. Article ID 932181 (2012)
13. D. Baimel, S. Tapuchi, N. Baimel, Smart grid communication technologies—overview, research challenges and opportunities, in *2016 International Symposium on Power Electronics, Electrical Drives, Automation and Motion (SPEEDAM)* (IEEE, 2016), pp. 116–120
14. A. Kovács, R. Bátai, B.C. Csáji, P. Dudás, B. Háy, G. Pedone, J. Váncza, Intelligent control for energy-positive street lighting. Energy **114**, 40–51 (2016)
15. S. Kakran, S. Chanana, Smart operations of smart grids integrated with distributed generation: a review. Renew. Sustain. Energy Rev. **81**, 524–535 (2018)
16. A. Stefanov, C.C. Liu, Cyber-power system security in a smart grid environment, in *2012 IEEE PES Innovative Smart Grid Technologies (ISGT)* (IEEE, 2012), pp. 1–3

17. A. George, A. Ravindran, M. Mendieta, H. Tabkhi, Mez: an adaptive messaging system for latency-sensitive multi-camera machine vision at the IoT edge. IEEE Access **9**, 21457–21473 (2021)
18. N.M. Thomasian, E.Y. Adashi, Cybersecurity in the internet of medical things. Health Policy Technol. **10**(3), 100549 (2021)
19. A.O. Otuoze, M.W. Mustafa, R.M. Larik, Smart grids security challenges: classification by sources of threats. J. Electr. Syst. Inf. Technol. **5**(3), 468–483 (2018)
20. P. Singh, M. Masud, M.S. Hossain, A. Kaur, Blockchain and homomorphic encryption-based privacy-preserving data aggregation model in smart grid. Comput. Electr. Eng. **93**, 107209 (2021)
21. R.C. Green, L. Wang, M. Alam, Applications and trends of high-performance computing for electric power systems: focusing on smart grid. IEEE Trans. Smart Grid **4**(2), 922–931 (2013)
22. J. Baek, Q.H. Vu, J.K. Liu, X. Huang, Y. Xiang, A secure cloud computing based framework for big data information management of smart grid. IEEE Trans. Cloud Comput. **3**(2), 233–244 (2014)
23. F. Wu, L. Xu, X. Li, S. Kumari, M. Karuppiah, M.S. Obaidat, A lightweight and provably secure key agreement system for a smart grid with elliptic curve cryptography. IEEE Syst. J. **13**(3), 2830–2838 (2018)
24. A. Ghosal, M. Conti, Key management systems for smart grid advanced metering infrastructure: a survey. IEEE Commun. Surv. Tutor. **21**(3), 2831–2848 (2019)
25. Z. Erkin, J.R. Troncoso-Pastoriza, R.L. Lagendijk, F. Pérez-González, Privacy-preserving data aggregation in smart metering systems: an overview. IEEE Signal Process. Mag. **30**(2), 75–86 (2013)
26. H. Bao, R. Lu, A lightweight data aggregation scheme achieving privacy preservation and data integrity with differential privacy and fault tolerance. Peer-to-Peer Netw. Appl. **10**, 106–121 (2017)
27. S. Zhao, F. Li, H. Li, R. Lu, S. Ren, H. Bao, S. Han, Smart and practical privacy-preserving data aggregation for fog-based smart grids. IEEE Trans. Inf. Forensics Secur. **16**, 521–536 (2020)

Chapter 16
Enhancing Packing Processes Through an Ergonomic Worker Study Using Digital Human Modeling and Environmental Assessment

Pratzanee Sungkaratn and Chootrakul Siripaiboon

Abstract The aim of this study is to improve the lifting process for 25 kg boxes vertically stacked on pallets, with consideration for the inclusion of female workers as per labor regulations. Utilizing the National Institute for Occupational Safety and Health (NIOSH) Composite Lifting Index (CLI) through Digital Human Modeling (DHM), the risk of musculoskeletal disorders (MSDs) from a single worker lifting, who performs 6 lifts per hour and works 8 h a day, was evaluated. The CLI for a single worker was determined to be 3.93, with recommend weight limit of 6.36 kg according to DHM. The study compared three lifting methods: Two-person manual lifting resulted in a CLI of 1.97, indicating a moderate risk of injury. Pre-leveling equipment before single lifting led to a CLI of 2.77, though this approach may extend lifting time. Lastly, spring balancer assistance resulted in a CLI of 0.47, indicating maximum safety but requiring investment in installation. In conclusion, the analysis suggests that process efficiency can be boosted by up to 55 lifts per hour, reducing cycle time and workers. Moreover, the carbon footprint of the facilities was calculated, indicating a reduction of 0.96 tons of CO_2 emissions per year compared to other options. This reduction significantly contributes to the establishment of a green manufacturing.

Keywords DHM · NIOSH · Lifting index · Ergonomics · Carbon emission

P. Sungkaratn · C. Siripaiboon (✉)
Division of Industrial Technology, School of Science and Technology, Sukhothai Thammathirat Open University, Nonthaburi 11120, Thailand
e-mail: chootrakulsiripaiboon@gmail.com

C. Areeprasert et al. (eds.), *Proceedings of the International Conference on Sustainable Energy and Environmental Technology for Circular Economy*, Springer Proceedings in Energy, https://doi.org/10.1007/978-981-96-2306-8_16

16.1 Introduction

Nowadays, a wide variety of new products are emerging due to innovative designs and increasing demands. Consequently, businesses need to store raw materials and finish goods in large quantities. Labor laws stipulate that female workers can lift a maximum weight of 25 kg, as specified in the Labor Protection Act of 1998, Sect. 37, However, data from 2022 shows that 763 individuals were injured from lifting or moving heavy objects, and 404 individuals were injured from work-related postures [1].

From literature review about MSDs found some study assessing the risk of body posture on repetitive and heavy lifting activities [2] that causes are related to awkward body postures, including bending, lifting, and physical exertion. The reviews have identified various research efforts to reduce work-related risks, such as adjusting the reach distances at noodle packaging stations in Korat [3] modifying work methods and designing equipment for lifting and moving squatting toilets [4] and improving workstations in an automotive parts factory in Chonburi. The DHM has been used to study ergonomics in various scenarios, such as concrete formwork activities in construction [5], university cafeteria workstations in China [6], The rest model of recovery time for 20 kg lifting task advice heart rate recovery time for 6 lifts per minute average 1.92 min [7].

Additionally, climate change is a global issue arising from increasing carbon emissions each year, partly due to the growing human population and various human activities. There were assessed 35 Thai students average CO_2 emissions approximately 2.16 tons of CO_2e per year [8]. The carbon emissions from improving manufacturing process need to be high considered.

This research was conducted to study the work process using a DHM to assess ergonomic risks according to the NIOSH lifting equation for lifting 25-kg product boxes, which is Thai female maximum the legal limit. The aim is to make informed decisions to reduce worker injuries and enhance process efficiency, while also minimizing waste of manufacturing, using the least amount of energy, and reducing carbon emissions.

16.2 Material and Methodology

16.2.1 Process Design and Improvement

To assess and improve the process, evaluating the Lifting Index (LI) is crucial for risk assessment using Tecnomatix software calculations. Initially, the default tasks in a packing room were observed (Fig. 16.1) to create DHM and design improvement processes for three options. Subsequently, the model was set for the default process (Fig. 16.2).

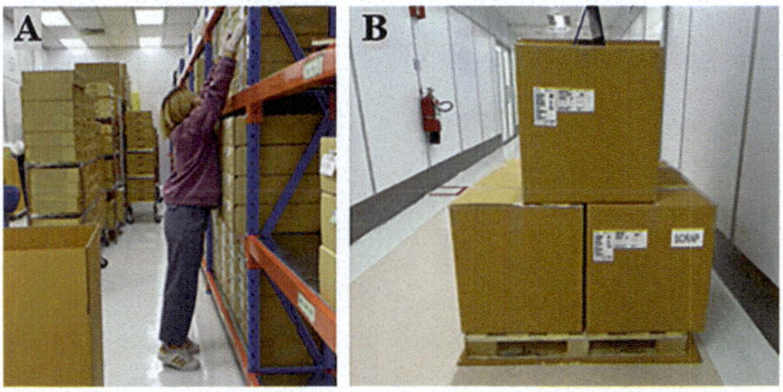

Fig. 16.1 Default task in packing room **a** Female working in process and **b** Stacked boxes on a pallet

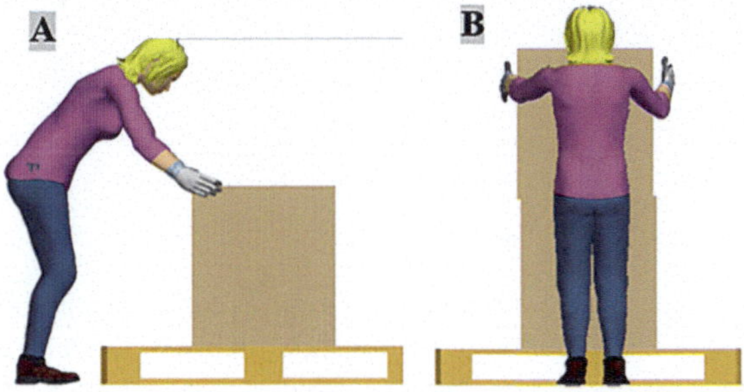

Fig. 16.2 The model set for default process **a** DHM place Box1 onto a pallet and **b** DHM stack Box2 over Box1

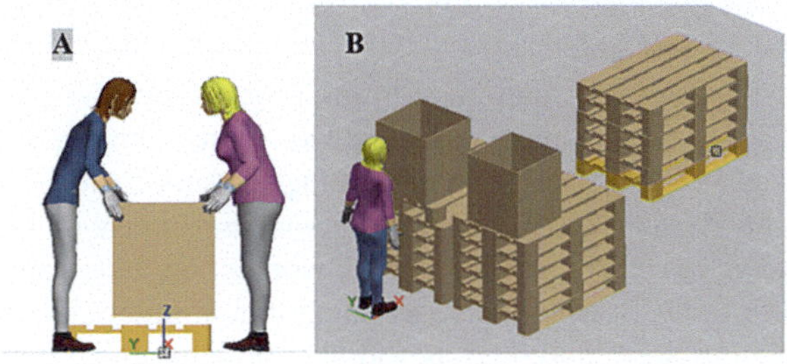

Fig. 16.3 The model set for options process **a** design option 1: 2 workers manual lifting and **b** design option 2: a system which adjustable height

The simulation report indicated that workers had CLI values exceeding 3, prompting the design and comparison of process improvements as follows: Option 1 involves assigning tasks to two workers for manual lifting without assistive tools and Option 2 implements a system where height adjustment precedes solo lifting (Fig. 16.3). Option 3 utilizes a spring balancer to aid in lifting the box (Fig. 16.4).

16.2.2 Simulation Procedure

In this study, the designed process was evaluated using a DHM with Siemens Tecnomatix Process Simulate Human software. The software utilized an anthropometric database for asian females and males to create accurate digital representations of the

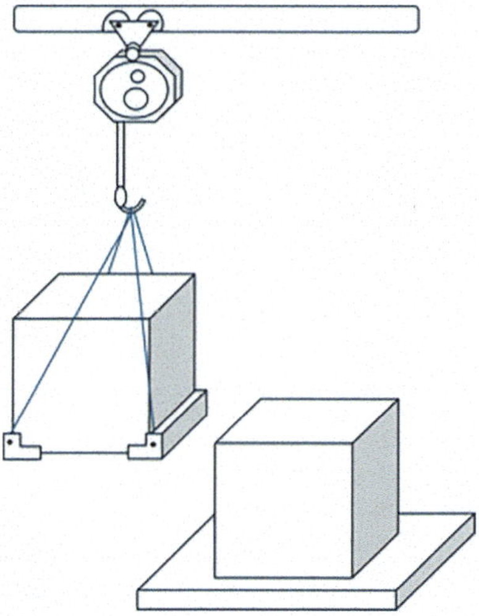

Fig. 16.4 Design option 3: The utilization of spring balancer lift the box

workers. These models were employed for ergonomic risk assessment and simulation of tasks involving lifting and moving product boxes within a factory's packaging room. Input parameters included task postures, a lifting frequency of 0.1 lifts per minute (6 lifts per hour), and an 8-h shift duration. Tools used to assess ergonomic risks included the NIOSH Lifting Index, which identifies tasks that may impact long-term safety.

The default equipment in the process includes product boxes measuring 530 × 530 × 600 mm and weighing 25 kg, along with wooden pallets of 100 mm height for transportation. DHMs representing workers at the 1st, 5th, 50th, and 95th percentiles of body sizes were created to encompass a range of worker physiques. These models served as representatives for the worker population and were used to simulate various lifting tasks, accurately assessing ergonomic risks across diverse body sizes. Task actions were assigned as follows: initially using both hands to grasp Box1 at the top position, lifting Box1 from the ground, and placing it onto a pallet at a height of 100 mm. Subsequently, moving to Box2's position to lift and vertically stack it over Box1 to generate ergonomic reports. This process was repeated for all required models, as illustrated in Fig. 16.5.

16.2.3 Ergonomic Risk Assessment

The DHM simulation results provide calculations for Lifting Index (LI), Recommended Weight Limit (RWL), Force Multiplier (FM), and Composite Lifting Index (CLI) using software, and report the results based on Eqs. 16.1 to 16.4, assessed from the NIOSH perspective and presented in Table 16.1. The goal in designing manual lifting tasks is to achieve an LI of 1.0 or less. LI values between 1 and 3 indicate

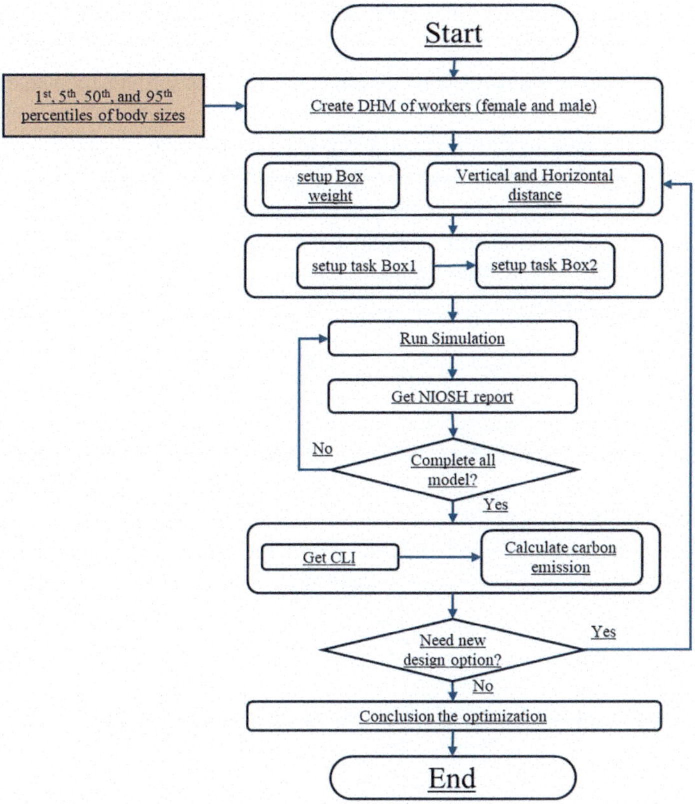

Fig. 16.5 Flowchart of simulation process using DHM

moderate risk, prompting considerations for process improvement, while LI values exceeding 3 indicate high ergonomic risk, necessitating immediate improvement.

LI is calculated using the following equation:

$$LI = \frac{L}{RWL} \tag{16.1}$$

where: **RWL (Recommended Weight Limit)** is the maximum weight, a worker can lift under ideal conditions.

Table 16.1 NIOSH criteria for risk levels of lifting index (LI) [9]

Lifting index	Criteria
LI < 1	Safe, low risk
1 ≤ LI < 3	Moderate risk; consider ergonomic improvements
LI ≥ 3	High risk: immediate ergonomic intervention needed

$$RWL = LC \times HM \times VM \times DM \times AM \times FM \times CM \quad (16.2)$$

LC is the Load Constant, follow by distance measurable with reduction coefficient for the equation.

HM is the horizontal multiplier, equal 1.0 if hand location is less than 250 mm.

VM is the vertical multiplier, the absolute value from optimum height 750 mm.

DM is the distance multiplier, is the travel distance of hands that computed.

AM is the asymmetric multiplier equal 1.0 when lifted directly in front of body.

FM is the frequency multiplier defined by frequency of lifting or duration of work follow by recovery time.

$$FM = \frac{RT}{WT} \quad (16.3)$$

CM is the coupling multiplier based on classification and vertical location of the lift.

L (Load): The actual weight of the load being lifted. (lbs. or kg).

Since this study is multi-task lifting, The Composite Lifting Index has been considered two identical boxes [9] as equation

$$CLI = STLI_1 + \sum \Delta LI \quad (16.4)$$

where STLI is computed for Single Task Lifting Index for each task

16.2.4 Carbon Emission

Once the simulation revealed results for the default process, it showed that options 1 and 2 pose ergonomic risks. The factory requires operators to work 8 h a day to transport 40 boxes. However, option 3, with the highest efficiency, can complete the task in just 0.72 h a day, resulting in the lowest electrical consumption for the 63-square-meter packing room. To calculate the greenhouse gas emissions equivalent to carbon dioxide (CO_2e) (Eq. 16.5), data collected from the factory indicates an annual electricity consumption of 18,900 kWh. The factor used is 0.5610 to convert this consumption to kg CO_2e per unit [10].

$$CO_2e\ Emission = Activity\ data \times Emission\ factor \quad (16.5)$$

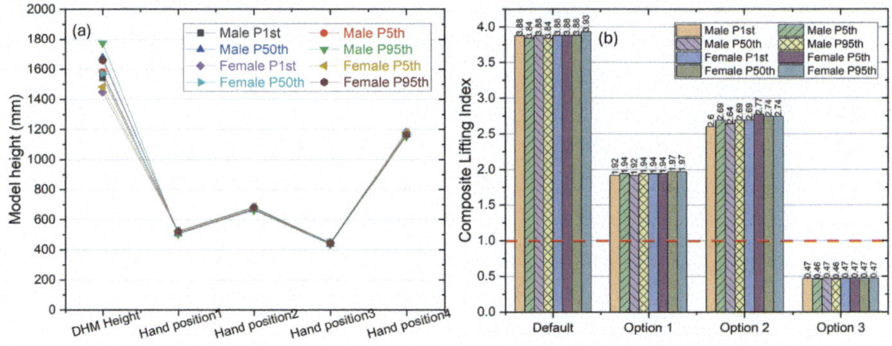

Fig. 16.6 **a** Comparison of hand positions during tasks and **b** Comparison of lifting index for each operator across different options

16.3 Result and Discussion

16.3.1 Process Simulation Results

In simulations involving 8 models with body heights ranging from 1449 to 1775 mm, hand positions were consistently placed on top of both boxes at the original position between 443 to 526 mm. However, during stacked tasks, the destination for Box2 was consistently higher than Box1, ranging from 1154 to 1184 mm. This indicates that the vertical hand position of workers is determined by the height of the box rather than being adjusted according to the workers' height (Fig. 16.6a).

Compares simulations across different genders and heights, showing that in the default process (Fig. 16.6b), a single worker had a maximum Lifting Index (LI) of 3.93, indicating a high risk of injury for all workers. In comparison, Option 1, which involved two-person manual lifting, resulted in a CLI of 1.97, indicating moderate risk. Pre-leveling equipment before single lifting reduced the CLI to 2.77. Lastly, using spring balancer assistance to reduce the load to 3 kg resulted in a CLI of 0.47, indicating maximum safety.

16.3.2 Carbon Emission Result

Carbon emissions compare with CLI across different processes designs (Fig. 16.7). The calculation shows 10.6 tons of CO_2e per 8-h workday annually for the default process, Option 1, and Option 2, all of which exceed an CLI of 1.0 despite a task frequency of just 0.1 lifts per minute. Option 3, achieving a safe CLI below 1.0, allows for increased efficiency by raising the frequency to 1.0 lift per minute, thereby completing the day's work in less than 8 h. This enhancement reduces electrical consumption to 0.96 tons of CO_2 per year.

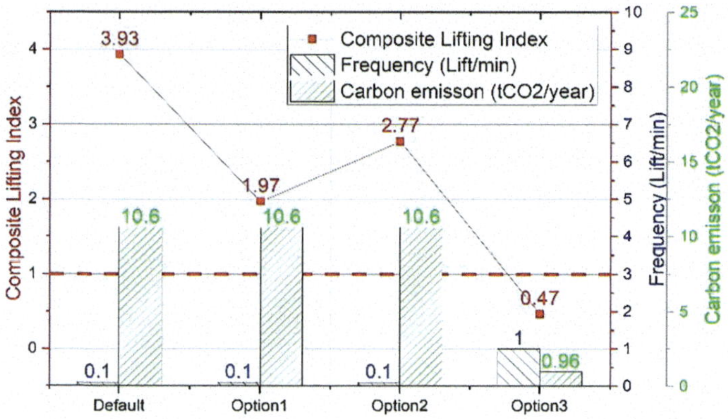

Fig. 16.7 Comparison of lifting index, frequency, and carbon emissions for each option

16.4 Conclusion

The risk of MSDs from lifting in the packing process is significant for workers performing 0.1 lifts per minute over an 8-h day, even when complying with regulations, especially when handling vertically stacked 25-kg boxes on pallets. This study recommends improving the process by option 3 which spring balancer assistance both lowest CLI and CO_2e emission. Analysis indicates that productivity can safely increase to a maximum of 1 lift per minute, reducing cycle times and the number of required workers. Furthermore, the carbon emissions for this enhanced process were calculated, revealing a potential reduction of 9.64 tons of CO_2e per year compared to other options. This approach allows the factory to prioritize both worker safety, as evaluated by Digital Human Models (DHM), and environmental sustainability by reducing carbon emissions from electrical consumption, contributing to its goals of green manufacturing practices.

References

1. The Office of Permanent Secretary Ministry of Labour, Thai labor statistics for 2022 12 (2022)
2. S.R. Kamat, N.M. Zula, N.S. Rayme, S. Shamsuddin, K. Husain, The ergonomics body posture on repetitive and heavy lifting activities of workers in aerospace manufacturing warehouse. IOP Conf. Ser. Mater. Sci. Eng. **210**(1), 012079 (2017)
3. P. Tantiwong, K. Suksai, N. Raksasombat, P. Duangtho, P. Jongkol, Application of reach distance data for improving workstation of Korat noodle packing. EAU Herit J. Sci. Technol. (Online) **16**(1), 114–126 (2022)
4. C. Maiton, N. Charoenchai, Risk assessment in lifting task in the squat toilet production. Thai Ind. Eng. Netw. J. **1**(1), 31–36 (2015)
5. V. Chellappa, J.S. Chauhan, Z. Strukova, Ergonomic risk assessment for concrete formwork activities in construction using digital human modeling, in International Ergonomics Conference (Springer, 2022)

6. J.W. Niu, X.H. Zheng, L. Zhang, S.Y. Xu, X. Li, S.S. Guo, S.T. Ding, et al., Investigation of ergonomics in Chinese university cafeterias' working situation at peak hours using jack, in 2011 IEEE 18th International Conference on Industrial Engineering and Engineering Management (IEEE 2011)
7. M. Amini, *Determination of Recovery Time for a Simple Lifting Task Based on Weight, Frequency, and Duration of the Lift* (Louisiana State University and Agricultural & Mechanical College 2017)
8. T. Utaraskul, Carbon footprint of environmental science students in Suan Sunandha Rajabhat University, Thailand. Proced. Soc. Behav. Sci. **197**, 1156–1160 (2015)
9. T.R. Waters, V. Putz-Anderson, A. Garg, Applications manual for the revised NIOSH lifting equation (1994)
10. I. Rodtusana, Assessment of Greenhouse Gas Emissions of Huachiew Chalermprakiet University (2013)

Chapter 17
Readiness Analysis for Implementing Green Diesel as Sustainable Aviation Fuels in Indonesia

Jeesica Hermayanti Pratama, Evtriyandani, Zeni Rahmawati, and Hamzah Fansuri

Abstract The rapid growth of the aviation industry has led to a rising demand for aviation fuel. This escalating reliance on fossil fuels raises concerns about the depletion of crude oil reserves and intensifying environmental consequences. As a response, alternative fuels are being developed, particularly environmentally sustainable and renewable biofuels. Among these, the conversion of biodiesel into green diesel has emerged as a promising solution. This conversion can improve fuel quality and address several issues associated with biodiesel, such as low energy content and incomplete combustion due to high oxygen content. Although green diesel policies have been in place in Indonesia since 2013, their practical implementation is still evolving. Sustainable aviation fuels (SAF) are essential for achieving zero-carbon aviation by 2050, requiring accelerated adoption through technological advancements and supportive policies. As the world's leading palm oil producer, Indonesia is well-positioned to lead the regional implementation of green diesel as SAF. Realizing this potential will depend on leveraging economic instruments, fostering inter-regional partnerships, and integration with international supply chains. This paper assesses Indonesia's readiness to adopt drop-in SAF from palm oil biodiesel.

Keywords Green diesel · Sustainable aviation fuel · Renewable energy

J. H. Pratama · Z. Rahmawati · H. Fansuri (✉)
Department of Chemistry, Faculty of Science and Data Analytics, Sepuluh Nopember Institute of Technology, Surabaya, Indonesia 60111
e-mail: h.fansuri@its.ac.id

J. H. Pratama
e-mail: jeesicahp@gmail.com; 6004221004@student.its.ac.id

Z. Rahmawati
e-mail: zeni.rahmawati@its.ac.id

Evtriyandani
Department of Environmental Engineering, Faculty of Civil, Planning and Geo Engineering, Sepuluh Nopember Institute of Technology, Surabaya, Indonesia 60111
e-mail: evtriyandani96@gmail.com; 6014222006@student.its.ac.id

C. Areeprasert et al. (eds.), *Proceedings of the International Conference on Sustainable Energy and Environmental Technology for Circular Economy*, Springer Proceedings in Energy, https://doi.org/10.1007/978-981-96-2306-8_17

17.1 Introduction

The escalating demand for aviation fuel is driven by the expanding commercialization of the aviation industry. As per the Global Outlook for Air Transport [1], the airline's fuel expenses in 2021 increased by approximately 30% as a result of the easing of travel restrictions and the initial recovery in global passenger demand. Aviation fuel is a significant cost for an airline, typically accounting for 20–25% of the total expenses. The proportion of overall expenses in 2022 is significantly higher than in 2021 as a result of the commencement of the Russia-Ukraine conflict, which led to a substantial surge in worldwide oil prices. As of June 2, 2022, the price of Brent crude oil stood at USD120 per barrel. A noteworthy feature of the fuel market this year is the substantial disparity in prices between crude oil and jet fuel. Presently, the increased fuel costs do not seem to be affecting individuals' inclination to travel. Nevertheless, once consumers have adjusted for their decreased travel, the diminished responsiveness of demand to price may decline in 2023, posing a challenge for airlines to manage the rise in fuel prices relative to demand in the upcoming year. Therefore, utilizing the more cost-effective aviation fuel will prove highly advantageous.

Air transportation necessitates the use of drop-in fuel that possesses a chemical structure compatible with the aircraft's built-in engine. Sustainable aviation fuel (SAF) or bio-jet fuel (bioavtur) can potentially decrease CO_2 emissions by 55% to 92%. SAF compability can substitute traditional kerosene in aircraft and fuel systems without requiring any adjustments [2]. Green diesel, classified as a second-generation biofuel, can be further processed to meet the stringent standards for aviation use, making it a potential feedstock for SAF production. Green diesel can be produced by upgrading biodiesel, the first-generation biofuel also known as fatty acid methyl ester (FAME) [3]. Green diesel can be produced by upgrading biodiesel, a first-generation biofuel commonly referred to as fatty acid methyl ester (FAME). Biodiesel, however, presents several limitations, including low thermal and oxidative stability due to its high oxygen content, reduced energy content, and the risk of carbon buildup in engines. Additionally, it is prone to degradation during storage. Converting biodiesel into green diesel enhances fuel quality and brings it closer to meeting SAF performance standards.

Extensive research has been conducted to support green diesel as an alternative biofuel, focusing on catalyst development, the use of highly variable feedstocks, and the conversion of biodiesel into green diesel. However, Indonesia's readiness to implement green diesel production technology and its application in SAF still requires further study. Therefore, this article was aimed to analyze Indonesia's readiness to implement green diesel in accordance with regulations and environmental ethics.

17.2 Materials and Methods

This article reviews literature data acquired from the Scopus database (https://www.scopus.com/) due to its high-quality criteria of indexation inclusivity. The Scopus database is known for its comprehensive collection of interdisciplinary academic papers sourced from reputable scientific journals. The database encompasses a diverse array of scientific disciplines, positioning it as a prominent repository for citations and abstracts. To analyze the regulatory frameworks, this article summarizes published national press release, report, news, and government policy.

17.3 Results and Discussion

17.3.1 Green Diesel Properties

Green diesel consists of a mixture of straight and branched chain-saturated hydrocarbons that typically contain 15 to 18 carbon atoms per molecule (C15 to C18). The composition resembling petroleum diesel allows the use of green diesel in compression ignition engines in pure form or as a mixture with the desired mixing ratio without the need for engine modifications. The suitability characteristics of green diesel as a fuel have met standards based on ASTM D-97, ASTM D-973, ASTM D-1298, and ASTM D4294–03, regarding flash point, energy content, cetane number, viscosity, density, and oxygen, sulfur and water content [4]. Green diesel also has good performance in reducing fuel consumption and NO_x emissions [5]. Green diesel has a higher cetane number above 70 compared to biodiesel so it can operate with higher thermal efficiency and lower fuel consumption. Using green diesel as a SAF has numerous benefits as it can also help to reduce the environmental impact of the aviation industry (Fig. 17.1). By 2050, biofuels are expected to reduce CO_2 emissions from the aviation industry by more than 50%.

17.3.2 Feedstock Availability

The physical and chemical characteristics of biodiesel can vary according to the raw materials that used. Palm oil has good oxidation stability compared to oils from other feedstocks [7]. It has almost the same amount of saturated and unsaturated fatty acids. Palm oil is composed of monoalkyl esters of long-chain fatty acids, primarily containing 17–19 carbon atoms, that can be transformed into biodiesel using a catalytic method. According to Li et al. [3], palm oil biodiesel content mainly consists of 17.42% methyl palmitate; 73.54% methyl oleate; and 7.26% methyl stearate. Palm oil biodiesel shows great potential as a viable alternative to diesel fuel.

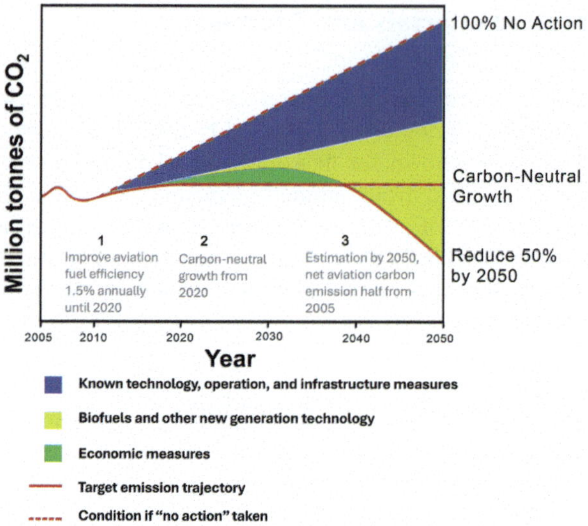

Fig. 17.1 CO_2 emission potential from aviation industry. Modified from Qaseem et al. [6]

As a major export commodity in Indonesia, palm oil ensures a stable supply of feedstock for biodiesel production. Palm oil is primarily used for food products (16.4%), oleochemicals (33.3%), biodiesel (14.1%), and exports (66.2%) in the country. Approximately 5 million tons of palm oil are available as a significant source for sustainable biodiesel production. Indonesia's production of palm oil biodiesel surged from 190 thousand kL in 2009 to nearly 8.4 million kL in 2019 [8]. Along with the government's strong commitment to creating an ecosystem climate conducive to the development of Indonesia's biodiesel industry, this has resulted in an increase in the biodiesel industry (also known as the biofuel business entity), both in terms of the number of companies and production capacity. Figure 17.2 displays the mapping of biodiesel producers in Indonesia. The rise in biodiesel production aligns with the Ministry of Energy and Mineral Resources' goal for the B-35 application target in Indonesia [9].

17.3.3 Technological Feasibility and Production Process

Generally, green diesel can be produced from vegetable oil through a deoxygenation (DO) reaction using supported metal catalysts [10]. The difficulty in determining the reaction stages and conditions of the DO reaction of vegetable oils is a factor in the massive development of fatty acid and fatty acid methyl ester (biodiesel) based feedstocks for green diesel production [5]. Several reaction pathways can involve simultaneously or individually in DO reaction, including hydrodeoxygenation (HDO), decarboxylation (DCO_2), and decarbonylation (DCO) [11]. The deoxygenation reaction is a thermal process of alkane production through the removal of

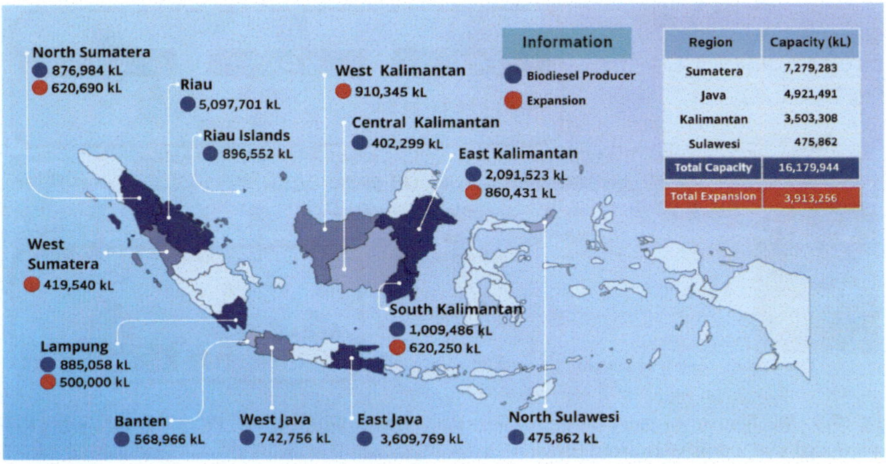

Fig. 17.2 Distribution capacity of biodiesel producers in Indonesia until 2022. Adapted data from APROBI (Indonesia Biofuel Producer Association)

hydrogen, oxygen and carbon atoms from fatty biomass by releasing CO_2, CO and H_2O. The involvement of the three different pathways in a deoxygenation reaction is strongly influenced by the catalyst and reaction conditions. The DO process employs a solid phase heterogeneous catalyst, which comprises an active phase deposited on a catalytic support material. The catalytic properties are influenced by the selection of the support and active phase, as well as the chemical interactions between the catalyst and the reagents and products. Nevertheless, when choosing the catalyst type, additional factors such as the ability to regenerate, the associated costs of regeneration, and the level of toxicity are also given due consideration [12]. Srifa et al. [13] reported that Co and Ni are the most effective catalysts for producing green diesel from palm oil. The proposed mechanism for this process can be seen in Fig. 17.3 revealing that stearic acid, octadecanol, octadecyl stearate, and stearyl oleate are oxygenated intermediates that result from the DO reaction. The proposed reaction mechanism involves the initial hydrogenation of double bonds in unsaturated triglycerides to saturated ones, followed by hydrogenolysis to form fatty acids and propane. The fatty acid undergoes DO at the metal catalyst site through DCO and HDO reaction pathways.

Green diesel research is still ongoing, particularly to target it as a bio-jet. From 1993 to 2022, there were 4294 publications (articles, proceeding paper, and review articles) with keywords of green diesel OR deoxygenation AND catalyst AND biojet AND biofuel. The term green diesel was continuously mentioned with particular trend as illustrated in Fig. 17.4. The annual number of green diesel article significantly improved anually. Indonesia can pursue further advancements in acquiring sufficient biojet fuel due to the abundance of publications pertaining to green diesel research. Furthermore, the presence of abundant raw materials offers significant

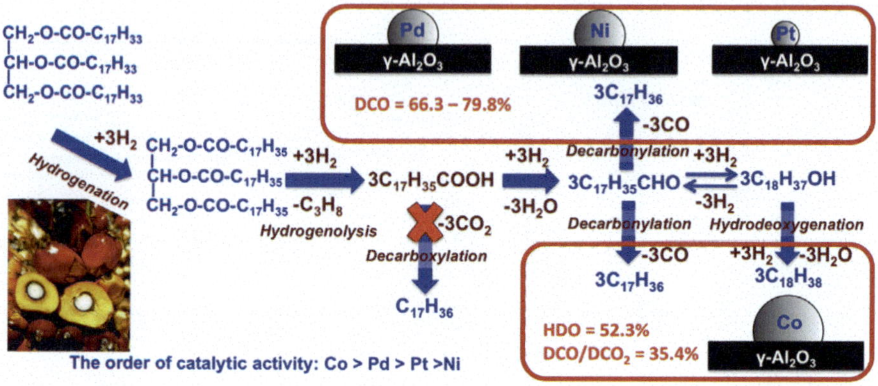

Fig. 17.3 Mechanism of palm oil deoxygenation reaction with Pd, Ni, Pt, or Co catalysts [13], reproduced with permission from Elsevier

research opportunities for acquiring appropriate catalysts and reaction conditions when implemented on a large-scale industrial level.

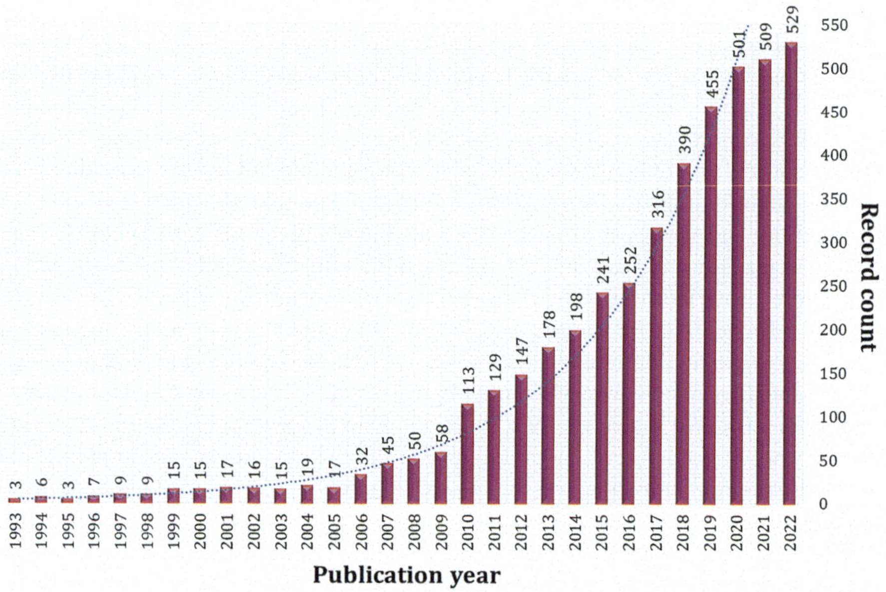

Fig. 17.4 Annual publications of the selected keywords from 1993 to 2022. (*Source* Authors' classification based on Scopus database)

17.3.4 Regulatory Framework and Development Progress in Indonesia

The international aviation sector, guided by the International Civil Aviation Organization (ICAO), has set ambitious goals to enhance fuel efficiency by 2% annually until 2050 and achieve carbon–neutral growth from 2020. In line with these global efforts, Indonesia is pushing for a new and renewable energy (NRE) mix target of 23% by 2025, with biofuels playing a crucial role. After the success of the B30 biodiesel program in land transportation, Indonesia has extended biofuel testing to the aviation sector.

A significant milestone was achieved in 2012 when Pertamina Research & Technology Innovation, in collaboration with the Center for Catalysis Engineering at the Bandung Institute of Technology, developed the Merah-Putih catalyst. This catalyst enabled the conversion of palm kernel oil into bioavtur feedstock. This innovation was further scaled up through a partnership with PT Kilang Pertamina Internasional, conducting industrial-scale production tests at the Cilacap Refinery. These tests successfully produced 2.4%-v bioavtur, known as J2.4, by co-processing Refined, Bleached, and Deodorized Palm Kernel Oil (RBDPKO) with kerosene using the Merah-Putih catalyst. In 2021, J2.4 bioavtur proved its superior performance in CN235-220 aircraft tests.

The Ministry of Energy and Mineral Resources Regulation No. 12 in 2015 mandates the blending of biofuels into aviation fuel, starting with 3% bioavtur by 2020, increasing to 5% by 2025. This regulation aligns with Sustainable Development Goals (SDGs) 7 and 13, which focus on ensuring access to affordable, reliable, and modern clean energy. To achieve these targets, the economic viability of J2.4 bioavtur must be supported through various government incentives, including tax benefits like super tax deductions for research and non-fiscal support. With Indonesia's estimated daily aviation fuel consumption of 14,000 kiloliters, the market potential for J2.4 bioavtur is projected to reach approximately 1.1 trillion rupiahs annually. This presents a significant opportunity for the growth of the national palm oil industry.

Indonesia has reached an important stage in the technological readiness to use green diesel as a sustainable aviation fuel. However, the success of future efforts will require the continued support of all stakeholders during the next phases of testing and commercialization. Establishing a robust aviation biofuel industry will depend on effective collaboration between the government, research institutions, bioavtur producers, and the aviation sector to bring the product to market successfully.

17.4 Conclusions

Green diesel as a sustainable aviation fuel (SAF) presents a highly promising alternative to conventional fossil fuels in the aviation sector. The production methods, materials used, and positive environmental impact make it a viable choice for reducing greenhouse gas emissions and improving air quality. The Indonesian government plays a pivotal role in fostering the development of SAF in the country. Indonesia has implemented a command-and-control strategy by imposing a mandatory requirement for blending SAF. However, it is important to note that there are currently no commercial flights in Indonesia that utilize bio-jet fuels in their daily operations. The government demonstrates its support through initiatives such as its participation in the pilot phase of the carbon offsetting reduction scheme for international aviation and the proposed establishment of a domestic carbon market known as the archipelagic carbon scheme. The development of SAF in Indonesia is driven by ongoing government policies and global support. Apart from challenges such as the limited existing research facilities, complex certification processes, low public awareness, and the need to develop catalyst performance, Indonesia has successfully achieved initial technological readiness by obtain D-100 and J2.4. The national project continues to advance green diesel for SAF implementation, with a target of achieving J5.0 by 2025. This ongoing development underscores the critical need for continued research, collaboration, and innovation to realize the full potential of green diesel in aviation, paving the way for a sustainable future.

Acknowledgements The first author gratefully acknowledges the Indonesian Endowment Fund for Education/Lembaga Pengelola Dana Pendidikan (LPDP), Ministry of Finance of the Republic of Indonesia for providing the financial support/studentship (Jeesica Hermayanti Pratama).

References

1. IATA: Global Outlook for Air Transport, https://www.iata.org/en/iata-repository/publications/economic-reports/global-outlook-for-air-transport---december-2022---presentation/. Accessed 16 July 2024
2. M. Song, X. Zhang, Y. Chen, Q. Zhang, L. Chen, J. Liu, L. Ma, Hydroprocessing of lipids: an effective production process for sustainable aviation fuel. Energy **283**, 129107 (2023)
3. X. Li, Y. Wu, Q. Wang, S. Li, Y. Ye, D. Wang, Z. Zheng, Effect of preparation method of NiMo/ γ-Al_2O_3 on the FAME hydrotreatment to produce C15–C18 alkanes. Renew. Energy **193**, 1–12 (2022)
4. J. Zafeiropoulos, G. Petropoulos, E. Kordouli, C. Kordulis, A. Lycourghiotis, K. Bourikas, Development of nickel catalysts supported on silica for green diesel production. Catal. Today **423**, 113952 (2023)
5. H.I. Mahdi, A. Bazargan, G. McKay, N.I.W. Azelee, L. Meili, Catalytic deoxygenation of palm oil and its residue in green diesel production: a current technological review. Chem. Eng. Res. Des. **174**, 158–187 (2021)
6. N.A.A. Qasem, A. Mourad, A. Abderrahmane, Z. Said, O. Younis, K. Guedri, L. Kolsi, A recent review of aviation fuels and sustainable aviation fuels. J. Therm. Anal. Calorim. **149**, 4287–4312 (2024)
7. R. Sakthivel, K. Ramesh, R. Purnachandran, P. Mohamed Shameer, A review on the properties, performance and emission aspects of the third-generation biodiesels. Renew. Sustain. Energy Rev. **82**, 2970–2992 (2018)
8. O. Farobie, E. Hartulistiyoso, Palm oil biodiesel as a renewable energy resource in Indonesia: current status and challenges. Bioenergy Res. **15**, 93–111 (2022)

9. T. Sipayung, Biodiesel Indonesia: Produksi, Konsumsi dan Ekspor (2023). https://palmoilina.asia/sawit-hub/biodiesel-indonesia/. Accessed 30 July 2024
10. J.H. Pratama, Z. Rahmawati, A.R. Widyanto, T. Gunawan, W.N.W. Abdullah, N.L.A. Jamari, A. Hamzah, H. Fansuri, Advancements in green diesel production for energy sustainability: a comprehensive bibliometric analysis. RSC Adv. **14**(48), 36040–36062 (2024)
11. Z. Rahmawati, L. Santoso, A. McCue, N.L. Azua Jamari, S.Y. Ninglasari, T. Gunawan, H. Fansuri, Selectivity of reaction pathways for green diesel production towards biojet fuel applications. RSC Adv. **13**, 13698–13714 (2023)
12. R.S.R.M. Hafriz, I. Nor Shafizah, N.A. Arifin, A.H. Maisarah, A. Salmiaton, A.H. Shamsuddin, Comparative, reusability and regeneration study of potassium oxide-based catalyst in deoxygenation reaction of WCO. Energy Convers. Manag. X. 13 (2022)
13. A. Srifa, K. Faungnawakij, V. Itthibenchapong, S. Assabumrungrat, Roles of monometallic catalysts in hydrodeoxygenation of palm oil to green diesel. Chem. Eng. J. **278**, 249–258 (2015)

Chapter 18
Life Cycle Carbon Emissions Analysis of a Detached House in Thailand

Nattaya Sangngamratsakul, Kuskana Kubaha, and Siriluk Chiarakorn

Abstract The high demand of fossil fuels used across various sectors leads to an increase of carbon emissions released into the atmosphere. This contributes to environmental issues such as global warming, air pollution, and climate change. Thailand's commitment to achieving carbon neutrality by 2050 highlights the critical need to assess carbon emissions in residential buildings. This is due to the significant contributions of both embodied and operational emissions from buildings to overall greenhouse gas (GHG) emissions throughout their lifespans. This study aims to determine the total carbon emissions of a detached house in Thailand by using the life cycle carbon emissions analysis (LCCEA) approach. The cradle-to-grave scope was employed for determining the total carbon emissions from six stages throughout the building's life cycle, i.e., initial, transportation, construction, recurrent, operational, and demolition. A detached house was broken down into 16 main building materials to quantify initial embodied emissions and recurrent embodied emission. The emission factors from the Inventory of Carbon & Energy (ICE) were employed to calculate initial embodied emissions and recurrent embodied emissions. The results from this study revealed that fiber-cement had the highest carbon emissions, accounting for 31.92% followed by steel (30.11%), and concrete (17.35%). The quantification of transportation emissions, construction emissions, operational emissions, and demolition emissions utilizes emission factors obtained from the Thailand Greenhouse Gas Management Organization (TGO). The results show that operational emissions account for the largest proportion at 54%, followed by initial embodied emissions (35%), recurrent embodied emissions (9%), and the remaining stages (2%). The total life cycle carbon emissions of a detached house were 53.97 $kgCO_2e/m^2/year$. These

N. Sangngamratsakul (✉) · K. Kubaha
Energy Management Technology Program, School of Energy, Environment and Materials, King Mongkut's University of Technology Thonburi, 126 Pracha Uthit Rd., Bang Mod, Thung Khru, Bangkok 10140, Thailand
e-mail: nattaya.s@mail.kmutt.ac.th

S. Chiarakorn
Environmental Technology Program, School of Energy, Environment and Materials, King Mongkut's University of Technology Thonburi, 126 Pracha Uthit Rd., Bang Mod, Thung Khru, Bangkok 10140, Thailand

C. Areeprasert et al. (eds.), *Proceedings of the International Conference on Sustainable Energy and Environmental Technology for Circular Economy*, Springer Proceedings in Energy, https://doi.org/10.1007/978-981-96-2306-8_18

findings can be used to develop guidelines for reducing carbon emissions in building construction sector, ultimately contributing to achieve carbon neutrality by 2050 and fostering a sustainable future society.

Keywords Life cycle carbon emissions analysis · Carbon emissions · Embodied carbon

18.1 Introduction

Rapid population growth is driving an increased demand for fossil fuels across various sectors and lead to a rise in carbon emissions, contributing to environmental issues such as global warming, air pollution, and climate change. Recognizing the urgency of this situation, Thailand announced the commitment for achieving carbon neutrality by 2050 at the 26th United Nations Climate Change Conference (COP26). According to a study, construction and operation of buildings consume 35% of global energy and are responsible for nearly 40% of carbon emissions [1]. Therefore, assessing carbon emissions in residential buildings is crucial for achieving carbon neutrality by 2050. Life cycle carbon emissions analysis (LCCEA) is a subset of life cycle assessment (LCA) that focuses specifically on determining the total GHG emissions associated with a building throughout its lifespan. The results are typically reported in kilograms of carbon dioxide equivalent ($kgCO_2e$) and usually account for the three main types of GHGs: carbon dioxide (CO_2), methane (CH_4), and nitrous oxide (N_2O). Carbon emissions can be categorized into two main types: embodied emissions and operational emissions [2]. Embodied emissions can be further divided into initial embodied emissions, which occur during the extraction of raw materials, transportation to the manufacturer and manufacturing of building materials. While, recurrent embodied emissions occur during maintenance and replacement of building materials throughout the building's lifespan. Moreover, embodied emissions occur during transportation of building materials to construction site, construction, and demolition process. Operational emissions, in contrast, refer to the emissions that occur during the building's use phase. This type of emission comes from running appliances, such as air-conditioning and lighting, to maintain comfortable conditions for occupants. The determination of carbon emissions through the LCCEA approach becomes increasingly important for assessing GHG emissions. This approach provides comprehensive results that reflect the environmental problems associated with climate change, a pressing global issue. This research applies LCCEA to determine the total carbon emissions of a detached house in Thailand. The cradle-to-grave scope is employed, including six stages, i.e., initial, transportation, construction, recurrent, operational, and demolition. To quantify initial and recurrent embodied emissions, a detached house was broken down into 16 main building materials. The emission factors for these carbon emissions were obtained from the Inventory of Carbon & Energy (ICE) [3]. While, the emission factors from the Thailand Greenhouse Gas Management Organization (TGO) [4] were then utilized to quantify transportation emissions, construction emissions, operational emissions, and demolition emissions. The findings from this research can be further applied to set guidelines for achieving carbon neutrality by 2050, ultimately contributing to a more sustainable future society.

18.2 Methodology

This research uses a two-story detached house designed for four occupants as a case study. The house was developed by the Department of Alternative Energy Development and Efficiency (DEDE) [5]. The house has an internal floor area of 127 m^2 and a gross floor area of 215 m^2. It is assumed to have a lifespan of 50 years. The LCCEA approach with cradle-to-grave scope is employed, including six stages, i.e., initial, transportation, construction, recurrent, operational, and demolition.

18.2.1 Initial and Recurrent Embodied Emissions

The quantity of each building material used is multiplied by emission factors obtained from the ICE to calculate initial embodied emissions. The number of replacements for each building material is then counted to quantify recurrent embodied emissions. Service lifetimes for each building material are obtained from previous publications [6, 7]. The quantity from each building materials, service lifetime, and emission factors are presented in Table 18.1.

18.2.2 Transportation Emissions

A transportation distance of 50 km is assumed. During the initial stage, building materials are transported by a six-wheeled truck with a maximum load capacity of 11 tons. Pickup trucks with a maximum load capacity of 1 ton are used for transporting building materials during the recurrent stage. Round-trip transportation is considered for calculating emissions. Emissions for the delivery trip are calculated based on a 100% load, while the return trip emissions are based on a 0% load. A six-wheeled truck releases 0.0677 $kgCO_2e/tkm$ when fully loaded and 0.4273 $kgCO_2e/km$ when empty. In comparison, a pickup truck releases 0.1411 $kgCO_2e/tkm$ when fully loaded and 0.3131 $kgCO_2e/km$ when empty. The emission factors for this calculation are obtained from the TGO [4].

Table 18.1 The quantity, service lifetime, and emission factors of building materials

No	Building materials	Quantity used at each stage (kg)		Service lifetime [1]	Emission factors [2]
		Initial	Recurrent	(year)	($kgCo_2e/kg$)
1	Concrete	255,982	–	–	0.103
2	Steel	15,157	–	–	3.02
3	Autoclaved aerated blocks	7120	–	–	0.280
4	Cements	7029	–	–	0.832
5	Sand	5745	–	–	0.00747
6	Aluminum	795	–	–	6.67
7	Glass	604	–	–	1.44
8	Fiber-cement panels	18,953	18,953	30	1.28
9	Timber	10,097	165	25	0.493
10	Ceramic tiles	4331	4331	25	0.780
11	Gypsum board	1627	1627	30	0.39
12	Plywood (MDF)	349	349	25	0.856
13	Fiberglass	271	271	15	1.35
14	Paints	133	102	10	2.91
15	Plastics	80	80	25	3.31
16	Plywood	32	32	25	0.681

[1] Based on Treloar [6] and Crawford [7]. [2] Based on ICE [3]

18.2.3 Construction Emissions

The energy consumption of the construction process was obtained from the study by Sangngamratsakul [8], which applied life cycle energy analysis (LCEA) to quantify the total energy consumption of an energy-conservation house in Thailand. This data was then used to determine construction emissions by allocating it evenly between diesel combustion and electricity. Emission factors for diesel combustion and electricity were obtained from TGO [4] at 2.7406 $kgCO_2e/L$ and 0.5986 $kgCO_2e/kWh$, respectively.

18.2.4 Operational Emissions

The operational energy data was obtained from Nitsunkit [9]. This data was then multiplied by the emission factor for electricity, which was 0.5986 $kgCO_2e/kWh$, as reported by TGO [4].

18.2.5 Demolition Emissions

Demolition energy consumption is assumed to be 176 MJ/m^2 based on Guan [10]. This value is multiplied by the gross floor area of the case study house to determine the total energy consumed during demolition. Assuming all this energy is derived from diesel combustion, demolition emissions are calculated using the emission factor for diesel combustion of 2.7406 kgCO_2e/L, as reported by TGO [4].

18.3 Results and Discussions

18.3.1 Embodied Emissions of Building Materials

Embodied emissions of the 16 main building materials used in a case study house were quantified for both initial and recurrent stages. Figure 18.1 presents the results of embodied emissions for each building material. The results reveal that fiber-cement releases the most carbon emissions, accounting for 31.92% of the total, followed by steel (30.11%), and concrete (17.35%). Notably, even though concrete is used in the highest quantity, fiber-cement and steel have a greater environmental impact due to their higher embodied emissions. Additionally, the results show that a small amount of aluminum can contribute a significant proportion of embodied emissions. These findings highlight the importance of material selection in reducing embodied emissions for buildings. Using materials with longer service lifetimes and greater durability can also reduce recurrent embodied emissions.

Fig. 18.1 The proportion of embodied emissions from different building materials

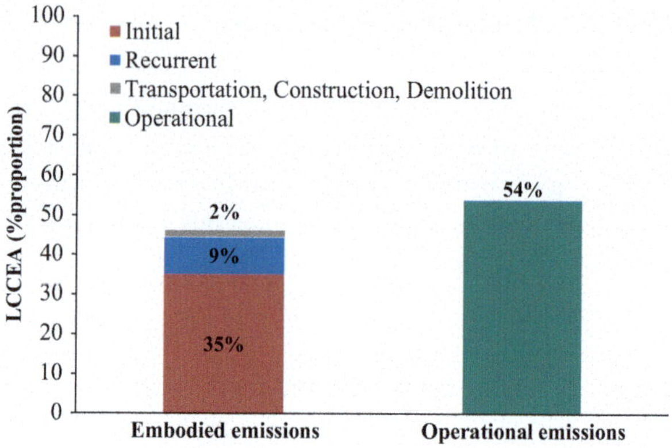

Fig. 18.2 The proportion of emissions throughout the building's life cycle

18.3.2 Life Cycle Carbon Emissions of a Case Study House

The total life cycle carbon emissions of a case study house are categorized into six stages: initial (120,786 kgCO_2e), transportation (1,347 kgCO_2e), construction (1,482 kgCO_2e), recurrent (31,224 kgCO_2e), operational (185,017 kgCO_2e), and demolition (2,847 kgCO_2e). As shown in Fig. 18.2, operational emissions contribute the most at 54%, followed by initial embodied emissions (35%), recurrent embodied emissions (9%), and the remaining stages (2%) comprise with transportation, construction, and demolition emissions. The total life cycle emissions of a case study house were 53.97 kgCO_2e/m^2/year. These findings emphasize that solely focusing on operational emissions is inadequate for reducing carbon emissions. Embodied emissions from both initial and recurrent stages are growing in importance. This study's results can inform the development of guidelines to reduce carbon emissions, ultimately contributing to achieve carbon neutrality by 2050 and promoting the development of a sustainable society.

18.4 Conclusion

These findings emphasize that reducing GHG emissions by focusing on operational emissions is inadequate. Embodied emissions from both initial and recurrent stages are growing in importance. In the context of Thailand, LCCEA throughout the building's life cycle should be emphasized and published to the public due to its crucial role in achieving the carbon neutrality goal by 2050. This study highlights the importance of selecting building materials with low embodied emissions, longer service lifetimes, and greater durability. These characteristics can significantly reduce both initial and recurrent embodied emissions. Furthermore, the results provide valuable data for future research on LCCEA of buildings. Ultimately, this research can inform the development of guidelines for achieving carbon neutrality by 2050 and supporting the development of a sustainable society for the future.

Acknowledgements The authors gratefully acknowledge the School of Energy, Environment and Materials, King Mongkut's University of Technology Thonburi, for facilitating this study. We would also like to thank the Department of Alternative Energy Development and Efficiency (DEDE) and the Thailand Greenhouse Gas Management Organization (TGO) for their valuable information support.

References

1. K.C. Ajay Kumar, A. Ghimire, B. Adhikari, H.R. Pant, B. Thapa, B. Baral, Life cycle energy use and carbon emission of a modern single-family residential building in Nepal. Curr. Res. Environ. Sustain. **7**, 1–9 (2024)
2. Z. Huang, H. Zhou, Z. Miao, H. Tang, B. Lin, W. Zhuang, Life-cycle carbon emissions (LCCE) of buildings: implications, calculations, and reductions. Engineering **35**, 115–139 (2024)
3. Circular Ecology Homepage, https://circularecology.com. Accessed 01 March 2024
4. Thailand Greenhouse Gas Management Organization Homepage, https://thaicarbonlabel.tgo.or.th. Accessed 01 April 2024
5. Study on Energy Efficiency Criteria and Prototype Development for Residential Buildings Homepage, https://www.aad.kmitl.ac.th/dede/. Accessed 01 March 2024
6. G. Treloar, R. Fay, P.E.D. Love, U. Iyer-Raniga, Analysing the life-cycle energy of an Australian residential building and its householders. Build. Res. Inf. **28**(3), 184–195 (2000)
7. R.H. Crawford, I. Czerniakowski, R.J. Fuller, A comprehensive framework for assessing the life-cycle energy of building construction assemblies. Archit. Sci. Rev. **53**(3), 288–296 (2010)
8. N. Sangngamratsakul, K. Kubaha, S. Chiarakorn, Embodied energy coefficient quantification and implementation for an energy-conservative house in Thailand. Sustainability **16**, 4045 (2024)
9. S. Nitsunkit, *Embodied energy of energy-efficient houses in Thailand* (King Mongkut's University of Technology Thonburi., Bangkok Thailand, Master of Engineering, 2018)
10. L. Guan, M. Walmsely, G. Chen, Life cycle energy analysis of eight residential houses in Brisbane, Australia, in 9th International Symposium on Heating, Ventilation and Air Conditioning (ISHVAC) and the 3rd International Conference on Building Energy and Environment (COBEE), (Procedia Engineering, Tianjin China, 2015), pp. 653–661

The manufacturer's authorised representative in the EU is Springer Nature Customer Service Centre GmbH, Europaplatz 3, 69115 Heidelberg, Germany. If you have any concerns regarding our products, please contact ProductSafety@springernature.com

Printed and bound by CPI Group (UK) Ltd, Croydon, CR0 4YY
01/10/2026
02225945-0013